ENERGY MANAGEMENT SYSTEMS IN BUILDINGS

The Practical Lessons

ENERGY MANAGEMENT SYSTEMS IN BUILDINGS

THE PRACTICAL LESSONS

Philip Gardner

Energy Publications

First published in Britain in 1984 by Energy Publications

ISBN 0 905332 34 2

Energy Publications is the book publishing imprint of
Cambridge Information and Research Services Limited
P O Box 147
Grosvenor House
High Street
Newmarket CB8 9AL
Tel: (0638) 663030

Printed in Great Britain by
St Edmundsbury Press, Suffolk

FOREWORD

by William Macintyre
Director General, Energy Efficiency Office

The need for the efficient use of energy in all spheres of national life has become increasingly apparent as the real cost of energy has risen in recent years. The future level of energy prices is notoriously difficult to predict but the indications are that in the long term the cost of scarce fossil fuels is likely to rise in real terms. It is therefore important to ensure that all energy users—from the householder to the industrialist—are aware of and make full use of all appropriate cost effective measures for improving energy efficiency.

The government's energy efficiency effort is spearheaded by the Secretary of State for Energy supported by the Energy Efficiency Office of the Department of Energy. The role of government in helping to bring necessary improvements about is clear: first to create a framework in which energy prices to the consumer reflect the true cost of production thereby discouraging profligate consumption; second, to ensure that information and advice is provided to enable appropriate efficiency improvement opportunities to be identified and wise investment decisions made; and third, to help the energy conservation industry promote its goods and services.

Three important services provided by the Energy Efficiency Office to non-domestic energy users are support for Regional Energy Managers Groups for exchange of experience and information; support for the development of energy monitoring and targetting management techniques; and the generation of independent information on newly available technologies through the Energy Conservation Demonstration Projects Scheme.

Electronic energy management systems for non-domestic buildings have been in the forefront of these new technologies and the Energy Efficiency Office has been particularly pleased by the progress being made in exploiting their potential. Indeed, the UK appears to have obtained a clear lead in the development and uptake of the latest techniques. It is particularly gratifying to see so many innovative UK equipment manufacturers applying advanced microprocessors to their products. Energy management systems have also been one of the most prominent and successful technologies featured in the Demonstration Projects Scheme.

The publication of this book is therefore timely; practical experience and quantitative information is now emerging from the demonstration projects and user interest is high. There is a need for guidance to enable prospective users to select the most appropriate system and obtain the maximum benefits from their investments. Dr Gardner's book encapsulates this experience and information in a manner that should be easily assimilable by a wide range of readers concerned with energy management. It follows on from a series of seminars and presentations he has given around the country in recent years.

Emphasis has been placed on practical lessons and advice rather than on detailed technical information and will be of value to those concerned with financial decisions as well as to those involved in the day-to-day management of energy.

This book will help disseminate the results of the Energy Conservation Demonstration Projects Scheme and thereby assist in achieving the maximum uptake of cost-effective electronic energy management technology in the UK. It will thus contribute to the eventual reduction of the burden of energy costs in our non-domestic buildings. It will also strengthen the credibility of the equipment manufacturers and help establish a new industry poised to exploit the substantial market that exists both at home and abroad.

CONTENTS

ILLUSTRATIONS

THE AUTHOR AND ACKNOWLEDGEMENTS

Philip Gardner was born in London in 1944. He graduated in physics from the University of Sussex in 1966 and, after a brief research career at the National Physical Laboratory, went overseas to undertake post-graduate studies in nuclear physics in 1968.

He obtained an MSc degree from McMaster University, Hamilton, Ontario, Canada in 1970 and a PhD from the Australian National University at Canberra in 1974.

On his return to England in 1974 Dr Gardner joined the Atomic Energy Research Establishment at Harwell. His first five years were spent dealing with technical and scientific problems arising from the commercial production of radioactive sources for medicine, industry and research. This was for an offshoot of the United Kingdom Atomic Energy Authority that was eventually privatised in 1982 as Amersham International.

In 1979 Dr Gardner moved to the Energy Technology Support Unit at Harwell. This was set up in 1974 mainly to provide project management expertise to the Department of Energy in the fields of renewable energy research, and research, development and demonstration of new technology for energy efficiency improvement. His principal involvement has been with the Department's Energy Conservation Demonstration Projects Scheme and his special responsibility has been electronic energy management systems for non-domestic buildings.

During this time, Dr Gardner has been in a unique position to develop a broad overview of the subject through 25 projects demonstrating various aspects of this new and exciting technology. Many problems have been encountered and much practical experience has been gained by equipment suppliers and users. His purpose in writing this book is to make this experience available to the widest possible audience so that maximum benefit may be obtained from the latest systems.

The author gratefully acknowledges the encouragement and support given him by Dr W M Currie Program Manager, Energy Conservation Demonstration Projects Scheme (ECDPS) at ETSU, the Energy Technology Support Unit, the United Kingdom Atomic Energy Authority (UKAEA), and the Department of Energy towards the publication of this book.

In addition, the constructive comments and suggestions given by his colleagues at ETSU have been extremely valuable and he is particularly grateful to Dr M J Butson for his generous assistance.

It should be noted that the views expressed in the book are those of the author and do not necessarily correspond with official government policy or opinion, or with information published under the ECDPS.

Chapter 1

INTRODUCTION — SAVING ENERGY IN BUILDINGS

Until a decade ago, most decisions about building services concerned provision of adequate comfort conditions at minimum capital cost. Energy was relatively cheap, so the optimum investment solution often involved practices that were very inefficient in terms of energy utilisation. We are now having to live with the consequences of these previous decisions at a time when energy costs have escalated in real terms and are likely to continue to do so for the foreseeable future. As a result, building services engineers, chief architects, estates managers, i.e. all those charged with the task of reducing energy costs in buildings under their care, are having to exercise considerable technical skill and judgement in identifying those efficiency improvement measures that are appropriate to their particular circumstances. They also need to ensure that those measures selected are properly implemented and utilised. It is proving no easy matter to rectify the mistakes of the past.

There are now many ways to improve the energy efficiency of existing buildings, some involving the way buildings are run, others requiring capital investment in various possible retrofit measures. Many of these improvement measures involve familiar, recognisable techniques such as insulation, higher efficiency plant, etc. Others concern new advanced technologies that may be unfamiliar and even bewildering to the older, traditionally trained buildings engineer. In the forefront of these new opportunities has been the exploitation of computer and microprocessor techniques to provide improved monitoring and control of building services. Whilst these techniques have become well established in other areas of energy use, e.g. in manufacturing industry for process control, they have been breaking entirely new ground in recent years as far as buildings are concerned. Consequently, it has been more difficult for the buildings people to get to grips with the technology, make the right investment decisions and get the most out of the equipment that is purchased.

The intention of this book, therefore, is to provide prospective users of electronic energy management systems in buildings with some useful practical information and advice. Hopefully, this will help them to decide whether or not to invest in the technology and, if they go ahead, to select the most appropriate system and to make the fullest possible use of it. Emphasis has been placed, therefore, on the lessons arising from existing projects, rather than on technical theory and detail. The latter has been given excellent coverage in a recent book by Abba Consultants (Automation) Ltd (Computer-based Energy Management in Buildings—published by Pitmans, 1982). Readers might find it useful to study these two books in conjunction.

Much of the material for the book has come from the author's involvement in a government scheme designed to help overcome some of the difficulties that may arise when investment decisions need to be made about using relatively new technology for energy efficiency improvement measures. In particular, there may be a lack of information on how the technology is likely to perform and consequently, the prospective user may be taking special risks by becoming involved with it. He may end up losing far more in terms of production or alienation of staff then he can expect to gain from reduced energy costs.

However, it is important that some pioneers do try out new technologies and the Department of Energy set up the Energy Conservation Demonstration Projects Scheme (ECDPS) in 1978 to encourage them to do so by sharing some of the financial risks. The scheme is also designed to provide for independent monitoring of the demonstration projects and to ensure that the results and general information arising from them are fully disseminated to the target market audience. Briefly, what the scheme offers is:

–a discretionary non-repayable grant of up to 25 per cent of capital, installation and other costs for selected projects meeting the full criteria for support

–the full costs of an appropriate monitoring exercise by an independent consultant

–the widespread publication and dissemination of results to the respective target audience.

In return, the organisation hosting the demonstration undertakes to provide all reasonable access to the project for the purposes of monitoring, and for interested parties who may be considering similar measures.

The scheme is managed on behalf of the Department of Energy by its Energy Technology Support Unit (ETSU) located at the Atomic Energy Research Establishment at Harwell. Fuller details of the scheme are provided in Appendix II.

The ECDPS covers a wide range of energy efficiency improvement technologies aimed at all sectors of national energy use. At the time of writing (1984), about 240 demonstration projects had been approved for funding. The majority of these concern industrial processes and techniques such as waste heat recovery and the use of waste materials as fuel substitutes. However, buildings also form an important part of the scheme with over 50 projects in non-domestic buildings and about ten involving housing. The most important demonstration technology for non-domestic buildings so far has been electronic energy management systems which account for roughly half of all the building projects. It is the experiences arising from these energy management system projects that form the basis of this book.

The Potential for Energy-Saving through Improved Controls

Before going on to deal with the technology, it is worth considering the potential value of electronic energy management systems to the nation in terms of reduced energy consumption. Figure 1.1 shows that 312 mtce (million tonnes of coal equivalent) primary energy was consumed in the UK in 1982. By primary energy, I mean energy consumed in its most basic form, e.g. coal, oil, gas, uranium, etc in boilers and furnaces of all types, including electricity generating stations. 1.0 tce is approximately equivalent to 0.6 tonnes of crude oil, 250 therms or 1.93 cubic metres of natural gas, 27.7×10^{10} J or 2,400 kWh of delivered electricity (taking into account losses at the power station, etc).

Figure 1.1 also shows that buildings of all types, i.e. domestic and non-domestic, account for nearly half (150 mtce per annum) of this national consumption. In fact, consumption in the buildings sector exceeds that of all industrial processes (95 mtce per annum). The bulk of the energy consumed in the buildings sector is in domestic buildings (92 mtce per annum). However, it is the remaining consumption (58 mtce per annum) in non-domestic buildings sector, namely commercial, public authority and industrial properties that will concern us in this book, in particular the potential for reducing this consumption through improved monitoring and control techniques.

Figure 1.1: Primary energy consumption in the UK.

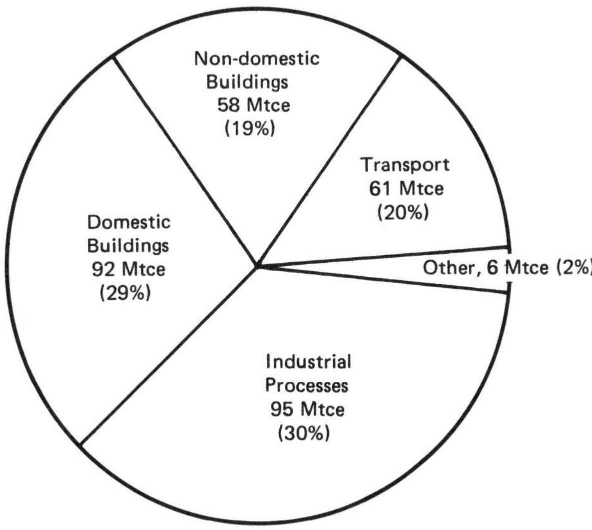

Figure 1.2: Breakdown of UK primary energy consumption by fuel use and building sector.

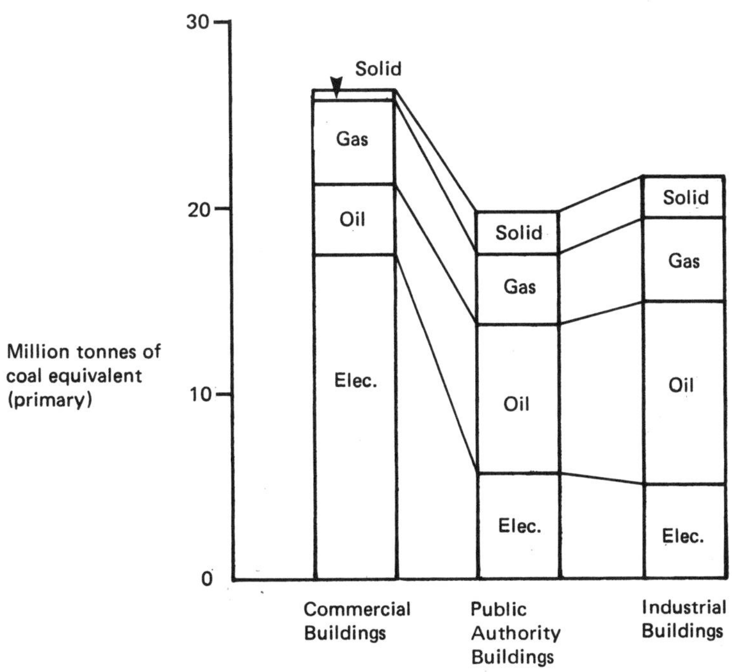

Figure 1.2 shows how this 58 mtce per annum primary energy consumption breaks down in terms of different fuels in the three main broad categories of non-domestic building. Oil is the major primary fuel in industrial and public authority buildings, with electricity accounting for only about one quarter of the total. This reflects the dominance of space heating requirements in such buildings. However, in commercial buildings, electricity accounts for about two thirds of the total consumption, probably reflecting the greater proportion used for lighting and air-conditioning.

It is recognised nowadays that there is substantial scope for reducing this national energy bill for buildings. This is not surprising when one considers that the vast majority were built before the recent start of the era of high-cost energy, at a time when it was cheaper to waste energy rather than provide the means to use it efficiently. The worst examples can often be found in air-conditioned office buildings where simultaneous heating and cooling is often applied in order to maintain required comfort conditions. The most dramatic savings would be made, therefore, by knocking down all our existing buildings and replacing them with the latest 'low-energy' designs, incorporating features such as:

—high standards of fabric insulation and draught-proofing

—mechanical ventilation with heat recovery

—attention to building orientation, shape and glazing to optimise the availability of passive solar heating and natural daylight

—use of high efficiency plant (e.g. boiler, heating and lighting, etc) carefully matched to the building requirements and minimising the need for artificial cooling

—advanced monitoring and control systems of the types to be described later in this book.

If we were to carry out such a complete rebuild, it should be possible to reduce national energy consumption for all buildings by at least 50 per cent or 75 mtce per annum. Clearly, this is not going to happen and, in any case would not be desirable for cultural and economic reasons. We are stuck with most of the buildings we already have, 80 per cent of which will probably still be in use by the end of the century. It is improvements in the way we run our buildings, combined with appropriate retro-fit technologies, therefore, that we are going to have to rely upon principally for future energy cost reductions. Advanced monitoring and control systems are going to be in the forefront of these opportunities.

Following chapters will show that full and proper exploitation of electronic energy management systems for improving present standards of monitoring and control should lead to savings of 10-30 per cent of the present energy consumption of an existing building. This will depend, of course, on present standards of operations, maintenance, control and good housekeeping. It will also depend on to what extent the user is able to make full and proper use of the technology. This, as we shall see, is not always very easy. If such systems were taken up in all non-domestic buildings, the technological potential for reducing the nation's energy bill will be between 6-18 mtce per annum.

However, the above figures ignore all the barriers that are likely to limit the actual uptake in the forseeable future. Many of these apply to all investment decisions on energy efficiency improvement measures, particularly where new technology is involved. Examples of such barriers are shown in Table 1.3.

The combined effect of all the barriers to investment described above is likely to limit severely the uptake of many energy efficiency improvement measures in buildings, particularly where new technologies are involved. Government measures may help to ease some of these barriers; in particular, the ECDPS and other publicity and promotion exercises will help to overcome those stemming from a lack of available information. Nevertheless, many barriers to investment will remain.

In the case of electronic energy management systems for buildings, we have the advantage of a lively market interest backed up by a vigorous and dynamic supply industry. Taking all factors into account, a reasonable target penetration would be about 20 per cent of the total market by the end of this century. An average fuel saving of 15 per cent per installation would then correspond to total

Table 1.3: Barriers to investment in improving energy efficiency

Technical and/or financial risk	The measure may not achieve the expected savings; even worse, it may lose more money than it is intended to save by interfering adversely with plant, processes or comfort conditions.
Lack of awareness	Many prospective users may not realise that the particular technologies are available.
Lack of available information	Many people might go ahead with an improvement measure if they could find more independent information on how it should be used and how it is likely to perform.
The high rate of return demanded from energy efficiency improvement measures	Industry, in particular demands a return on these investments which is often much higher than that demanded from other types of investment, e.g. in production facilities. It is commonplace to seek to recover the initial outlay in under 12 months.
The relative weakness of the equipment suppliers	Manufacturers of energy efficiency improvement technology are typically small, financially vulnerable, and ineffective in marketing and R & D compared to the major fuel suppliers, whose interests lie in increasing their share of the national energy market.

In addition, there are several barriers relating specifically to efficiency improvement measures in buildings, namely:

The long lifetimes of buildings	This tends to make the rate of introduction of new technologies slow compared to say industrial processes; most existing buildings have yet to adopt basic well-established measures such as time-switches and room thermostats for space heating control.
Behavioural factors	Many efficiency improvements will alter the comfort conditions of occupants, perhaps downwards if these were previously over-generous; care has to be taken not to annoy or irritate people.
Climatic factors	Whilst measures for a particular industrial process may be appropriate regardless of the location of any individual factory site, the same does not necessarily apply to buildings; what is appropriate in South Devon may not always be so in Caithness.
Traditional conservatism	Some energy saving measures in buildings can alter the visual appearance of the building, internally or externally. Matters of taste, construction practice, safety standards and regulations may inhibit their uptake. This will not apply to improved controls in many cases, but may prove a difficulty from time to time. For example, automatic lighting control in an office building may give rise to a more haphazard pattern or illumination as viewed from the street. This will represent good energy efficiency but may not necessarily appeal to the Chairman of the Board!

national energy savings of around three mtce per annum and a total capital investment of about £500 million at 1984 prices. Both of these sums are very substantial and well worth striving for, hence the government support of applications of the technology through the ECDPS. Indeed, the target national energy savings to be achieved through replication of the demonstrations of energy management systems are designed to eventually contribute some 0.8 mtce to the above amount and to stimulate investment of around £150 million.

At the moment, however, uptake of advanced electronic energy management technology for buildings is well under one per cent of the total potential market, although increasing rapidly as a result of the intensive efforts of the supply industry. Support has been given by government through information schemes like the ECDPS run by the Department of Energy and aimed at users and development schemes run by the Department of Trade and Industry to assist manufacturers. Nevertheless, many of the barriers to investment described above will persist and many users are not finding it easy to make full and proper use of the technology. The aim of this book, therefore, is to provide a little more information, assistance and guidance to existing and future users so that they may be in a better position to make wise investment decisions and to get the maximum benefit out of the systems that they purchase.

Chapter 2

BACKGROUND TO ENERGY MANAGEMENT TECHNOLOGY

What is an Energy Management System?

Some confusion appears to exist over the definition of the term 'energy management system'. In the UK, it might mean an 'all singing, all dancing' system for monitoring and controlling a multiplicity of building services. However, the term has also been used, particularly in the USA, to describe equipment which is principally concerned with shedding or cycling electrical loads (e.g. for maximum demand control). It is sometimes also applied to a wide range of other devices capable of controlling a single building service, such as heating or lighting. There is also the term 'building automation system' which is occasionally used as an alternative to 'energy management system'.

For the purposes of this book, the following definitions will apply:

–energy management technology can be defined as technology enabling the monitoring and control of building services on a closed loop basis such that energy savings provide the principal source of payback. Other financial returns might also be obtained, e.g. from savings of manpower, reduced maintenance, etc but energy savings predominate

–*a dedicated energy management system* is one which monitors and/or controls a single building service (such as lighting, heating, electrical load management, etc)

–*a comprehensive energy management system* is one which carries out a number or all of the energy monitoring and control functions that might be cost-effective in a particular building or site

–*a building automation system* is one which obtains a significant proportion of its payback from non-energy functions such as fire and security monitoring, document transmission, etc. Normally, this term would be reserved for very large scale systems installed on large sites.

What can Energy Management Systems Do?

Electronic energy management systems can save energy in a number of different ways which will be described in detail in Chapters 3 and 4. For example:

Automatically Switching Plant On and Off

Most people are aware from their personal experiences that enormous avoidable wastage of energy exists almost everywhere in industry, commerce and the home. One of the principal sources of such wastage arises from plant and machinery running unnecessarily during periods when it is not required. People are very good at switching things on when they think they need them but are not nearly so good at switching them off again afterwards. Considerable scope exists, therefore, for saving energy by automatically switching plant on and off (or just off) on the basis of time of day (to match working periods), type of day (to cater for weekends, etc) and environmental conditions (e.g. space temperatures, humidities, etc).

Switching things off wherever possible often represents the largest source of potential savings in a building and is a function for which electronic energy management technology is ideally suited. There are many possible applications; for example, lighting can be automatically switched to minimise use outside working hours, during mealbreaks, when occupants are away from their work stations or when ambient daylight levels are adequate. Additionally, heating systems can be controlled to prevent buildings being unnecessarily heated at night, during weekends, or to space temperatures above the statutory comfort limits.

Optimisation of Plant Operations and Services

The second most important area for saving energy through energy management technology is by improving the efficiency of plant operation. Good manual operations and maintenance practice can achieve sustained high levels of efficiency. However, accurate optimisation of all parameters of a complex building service needs the high speed computer power that can be provided nowadays so cheaply by the microprocessor.

A familiar example of sophisticated optimisation techniques is the adjustment of space heating start and stop times to take into account external temperatures and the thermal inertia of the building. The latest techniques enable this to be done on a self-learning, self-adaptive basis. Other examples are the optimisation of fuel/air mixture ratios for boilers (the so-called oxygen trim technique) and the maximised use of outdoor air for cooling in an air-conditioned building, such that the latent-plus-sensible heat load is minimised (enthalpy economisation).

Monitoring of Plant Status and Environmental Conditions

Further wastage of energy can occur by undetected malfunction of plant. For example, valves may jam, pipework may leak, controls may become mis-set or malfunction and give rise to excessive or deficient comfort conditions with a high probability of energy wastage. Modern microprocessor technology is able to collect information from sensors (e.g. temperature, pressure, flow, humidity, etc) and display readings in a suitable form to enable building services or maintenance engineers to readily assess the state of their plant. In particular, they can be quickly alerted to alarm conditions, e.g. when parameters drift outside pre-set limits, and thereby take more rapid remedial action than was previously possible. In this way, energy management technology can help to improve and preserve standards of operation and maintenance. An example of this type of feature can be found on the latest versions of oxygen trim systems for boilers. These can provide readings of flue gas temperatures and oxygen levels and of calculated combustion efficiencies at the touch of a button, on a digital display or regularly plotted on a chart recorder. Maintenance engineers can more easily maintain peak combustion efficiency by looking out for sudden changes which might indicate, for example, faults in the burners.

Provision of Energy Management Information

Energy cannot be effectively managed unless adequate information is available on which to base sound management decisions. Energy management technology can provide detailed and comprehensive information on energy flows and consumptions (from heat, steam, electricity and oil meters, etc), identify trends and help assess overall building energy performance. This type of information is likely to be of considerable value in the general task of energy management. For example, it could enable energy managers to assess the effectiveness of not only the energy management system itself, but also other subsequent energy efficiency improvement measures. It is also likely to be useful in assessing and providing evidence of acceptable comfort conditions. This may help in resisting pressure from inadequately clad occupants for more heat by demonstrating that the space temperatures are at or close to the statutory 19°C upper limit. This type of facility is a particular feature of comprehensive energy management systems.

Provision of a Remote Monitoring and Control Facility

Another of the important features of present comprehensive energy management technology is the scope for relaying information to and from remote outstations and a central supervisory station. These transmissions can be over internal cable networks in a single building or on a single site. They can also be over the public telephone network, using automatic dialling techniques.

On a large building or site, this enables a much more effective use of management and maintenance manpower. Instead of having to travel round and inspect individual items of plant or control systems, in order to check operations or take readings, etc, it is possible to use the computer peripherals of the energy management system central station (i.e. VDU, keyboards, printers, etc) to carry out the necessary actions remotely. The energy manager can now learn a great deal about what is going on in his buildings and execute certain actions without leaving his office. For example, he can adjust time settings when British Summer Time starts or ends with minimal effort.

This type of central information facility is obviously of particular importance where groups of buildings dispersed around the country are concerned. Savings over and above those due to improved local control may be achieved by energy managers being able to sustain correct settings and detect and rectify malfunctions much more effectively than with isolated dedicated controllers. For example, many local authorities have installed improved heating controllers in local schools. Substantial savings are expected and may be achieved in the first instance. However, these are gradually lost in many cases because, for example, local school staff interfere with the settings (perhaps to cater for special circumstances such as evening or weekend events) and fail to restore them to the correct values, or an undetected malfunction occurs. Either way, there is a risk of excessive fuel consumption (e.g. heating on 24 hours per day) and/or poor comfort conditions. A good well-trained school caretaker can often do much to maintain good standards of operations with simple plant, but more sophisticated equipment often has to wait for one of the infrequent regular maintenance visits before a defect is spotted and rectified.

A remote monitoring and control facility should, therefore, enable greater reliability of control system operations. Further benefits might be obtained from the effect that such a remote monitoring system could have on people's energy awareness and behaviour. For example, school occupants might be less likely to use window opening as a means of controlling room temperatures in winter if they were conscious of their actions possibly being instantly recorded in the council offices as a sudden drop in space temperature. Finally, there are the potential manpower savings that might arise from more effective use of the maintenance staff resources.

How do Energy Management Systems Save Energy?

In summary, therefore, the principal means by which energy management systems save energy are:

—by minimising the unnecessary use of plant

—by optimising the efficiency of plant

—by enabling improved standards of operations and maintenance

—by enabling improved standards of overall energy management

—by encouraging higher levels of energy awareness and good housekeeping.

In the following chapters, individual aspects of energy management technology will be discussed in more detail. Actual levels of achievable energy savings in buildings depend heavily on the particular function or functions being carried out by a system and on the underlying building circumstances. On average, however, it is reasonable to expect savings in the range 10-30 per cent of present consumption from full exploitation of the latest technology in typical circumstances. But, as we have already mentioned, such exploitation is not a simple matter and there are many pitfalls for the unwary prospective user.

Chapter 3

DEDICATED CONTROL SYSTEMS

A Simple or More Sophisticated System?

The first level of improved control in non-domestic buildings is likely to involve replacement of existing electromechanical controls by more advanced electronic (probably microprocessor-based) systems dedicated to single building services (such as heating, lighting, etc). Progress to comprehensive energy management systems might come at a later date.

This is not meant to imply that conventional electro-mechanical controls are inadequate or inappropriate in any way. Indeed, a substantial proportion of the savings achievable by switching plant off whenever possible can be obtained by the judicious application of simple time switches. Several organisations have embarked on an energy efficiency improvement campaign using a number of these types of devices in appropriate applications.

However useful simple time switches might be, they will still carry certain inherent disadvantages compared to more sophisticated systems. For example:

- microprocessor systems can provide much greater detail of programming than can simple electromechanical devices. Different programs can be provided for different types of day (e.g. workdays and weekends). There can be much greater numbers of start/stops per day. Different programs can also be provided for different zones of a building. A very large number of simple time switches would be required to match the programming sophistication of typical microprocessor-based systems
- solid state electronics is often much more reliable than its electromechanical equivalents
- microprocessor-based systems can often provide programmed control of a number of items of plant from a central point. This makes re-setting of programs considerably easier and quicker than is typically the case with conventional time switches. A simple adjustment to cater for a change from summer daylight saving time can entail many man-hours of effort if large numbers of clocks scattered around a large site are involved
- microprocessor control systems can also carry out certain optimisation functions (Chapter 2) that would not be possible with simple time switches

–microprocessor-based dedicated control systems are often compatible and assimilable with large-scale comprehensive energy management systems. This enables energy management technology to be approached in a modular way, if required; a user can start with improved dedicated controls in the first instance, linking them into a comprehensive energy management system at a later date.

Advanced dedicated control systems are now available for a wide range of building services. These will be described in more detail later in this chapter. Several of the categories of system have been the subject of ECDPS-supported demonstration projects.

Lighting Control

This is an especially important area for energy conservation for several reasons which will be discussed further in some detail. These include:

–the present very high levels of unnecessary usage of lighting that commonly occur in most types of buildings, particularly in office buildings

–the substantial proportion of total fuel costs that are often accounted for by lighting; for example, in office buildings, lighting can be responsible for up to half of the total energy bill, partly because of the relatively high costs of electricity.

Everyone is conscious of the fact that people are very good at switching services on when they feel they need them, but are remarkably inept at switching them off afterwards when they are no longer required. Lighting is no exception and patterns of typical switching behaviour have been observed in detail by scientists at the Building Research Establishment of the Department of the Environment.

What has been found is that people tend to assess the need for artificial lighting only at the beginning of a period of occupancy. This could be by cleaners in the darkness of early morning. They tend also to do this on the basis of the perceived minimum level of available natural daylight, for example, in the darkest corner of the space. Artificial lighting is often then kept at this initial level regardless of changes that might subsequently occur (such as the natural daylight increasing, people leaving the space for meal breaks, etc) until the end of the occupancy period (which might include cleaning or maintenance outside normal working hours).

This typical pattern of manual light switching gives rise to substantial unnecessary usage of lighting in three principal ways, namely:

–lights on when natural daylight levels are adequate

–lights on when people are away from their workstations

–lights on outside normal occupancy periods (e.g. during lunch breaks or cleaning periods, etc).

A typical set of circumstances is that commonly occurring in modern open-plan office buildings. The first arrivals (possibly cleaners or security staff), say on a winter's morning, turn on all the lights in their working areas by 'wiping' the rows of switches. Subsequent arrivals accept the levels of illumination provided because they appear to be more than adequate. However, no adjustment (such as switching some lights off) will occur to take account of increasing daylight levels in areas near windows as the morning progresses. Indeed, the typical arrangements of lighting circuitry often make the isolated switching of luminaires in window perimeter rows impossible.

Further unnecessary use will occur as individual workers leave their normal stations to attend to business elsewhere in or outside the building. This will peak at lunchtimes, when only a fraction of the staff are likely to be at their desks between about 1200 and 1400 hours.

Finally, at the end of the day (say 1700 hours), workers leave the building fully lit for cleaners and/or security staff to deal with. It is usually not until the last person leaves at say 2100-2200 hours that all but essential security lighting will be switched off. Thus, although the normal office day only extends for eight hours less meal breaks, with an actual maximum requirement for full lighting between about 0830 and 1030 and 1530 to 1730 (i.e. four hours), full lighting is likely to be provided from say 0700 to 2100 hours (i.e. 14 hours). This can represent up to 70 per cent wastage.

The widespread unnecessary use of lighting in office buildings can be easily observed by the passer-by. A casual inspection of any major city centre at night will reveal many brilliantly illuminated office blocks with no apparent occupancy. Other types of buildings (e.g. schools, factories, etc) will also display much wasteful use of lighting energy.

Actual levels of unnecessary use will depend on how the various spaces in the building are used. Schools, for example, have classrooms that may see regular changes of pupils and/or teachers. Under these circumstances, there is likely to be more frequent reassessment of lighting requirements than in the office sector, and consequently levels of actual wastage are likely to be somewhat lower, perhaps 10-20 per cent of the total.

There are, therefore, many different lighting circumstances and control requirements. Many of these have been studied by the Buildings Research Establishment. Briefly, these may be grouped into the following main categories:

- time-based switching on and off (i.e. to match overall working time schedules)
- time-based switching off only
- daylight sensing to take into account natural lighting availability
- occupancy sensing, turning lights off if spaces appear empty
- manual reset facilities to override any of the above control strategies, if required, to take the needs of individuals into account.

The above strategies may be combined in certain types of lighting control systems. Examples of these, drawn from projects supported under the Energy Conservation Demonstration Projects Scheme will now be described.

Time-Based Switching with Photocell Control of Perimeter Luminaires in Large Open-Plan Office Spaces

This type of strategy is particularly useful for large areas that have well defined work schedules. By using automation techniques to switch on in the morning, the wastage that occurs by manual switching during an extended security patrol or cleaning schedule prior to the start of the normal day can be minimised. More importantly, savings can also be made by ensuring that lights are not unnecessarily left on at the end of the working day. It is essential, however, to ensure that reasonable provision for local manual switching is retained. Certainly, lighting needs to be available, if required, to cater for abnormal working schedules, such as overtime. In addition, it is useful to retain a manual facility to switch lights off. Furthermore, for safety reasons, it may be necessary to retain conventional switching arrangements for lights in certain critical areas such as corridors and stairwells, etc.

Several technical options are now available to effect this type of switching strategy. The example which will be described here is drawn from a demonstration project in the 24,400 sq metre civic offices of Portsmouth City Council (Figure 3.1). The building design is very attractive but not very energy-efficient. The lighting load (60 W per sq metre) is especially high in this building because of the reduction in daylight availability caused by the use of tinted window glass and the attenuation of some of the artificial light by the false diffuser ceiling shown in Figure 3.2. Prior to the installation of the new control system, lighting costs were running at around £75,000 per annum. The subsequent need for lighting control stems from a requirement to reduce the lighting costs down to more reasonable levels. A lesson here for architects is to give more consideration to the operating costs of their designs at an early stage of a project.

THE NEW CIVIC OFFICES OF PORTSMOUTH CITY COUNCIL

Figure 3.1: The extensive use of tinted glazing contributes to a need for high artificial lighting levels.

Figure 3.2: An interior view—the attractive false diffuser ceiling also increases the necessary installed artificial lighting load.

Source: ETSU

Portsmouth City Council selected an ECS 500 lighting control system by ECS Energy Conservation Systems Ltd of Feltham, Middlesex. The system comprises central time switches (Figure 3.3) driving relays on the individual distribution circuits. An electromechanical time programmer is used in the Portsmouth project, but more recent versions exploit the enhanced capabilities provided by the microprocessor. The lighting distribution circuits conveniently correspond to the rows of luminaires in the large open-plan office spaces (Figure 3.2). Lights are turned on at 0800 hours and off again at 1730 hours. Push button reset panels (Figure 3.4) located at convenient central points on each floor allow individual rows to be turned back on again outside the above hours, if required. Subsequent off-signals are sent at regular intervals (0.5-2.0 hours) to turn out again any lights that may have been reset.

In addition to the time-based switching described above, the ECS 500 system can also provide photoelectrically controlled switching of rows of perimeter lights near to windows. Photocells mounted in positions just inside the windows measure the availability of ambient daylight and cause lights to be turned off if threshold levels are exceeded. An adjustable time delay allows brief changes due to passing clouds to be ignored, thus avoiding excessive frequency of switching.

This project has been independently monitored over a two-year period covering before and after the commissioning of the system. The report (which is freely available from ETSU–see Appendix II for details) shows that the initial estimates of 31 per cent savings were conservative and that actual savings are currently around 39 per cent giving a payback of just over two years on the original £53,000 investment. However, a longer payback period would have been achieved if the installed lighting load of 60 W per sq metre had been closer to more typical levels of around 30 W per sq metre.

A separate behavioural study has shown that lighting control systems of this type are acceptable to building occupants. However, lessons drawn from the study indicate that it is particularly important to ensure that all occupants are fully informed about the system well before its commissioning so that they clearly understand what its purpose is, why it is being installed and how they should use it. Some complaints were received during the initial teething period with a substantial proportion concerning the operation of the photocell control of perimeter luminaires. This needs to be carefully 'tuned' to avoid excessive frequency of switching which can be irritating to nearby occupants. However, overall the level of satisfaction with this installation has been high and it is generally regarded as a success.

Time-Based Switching of Lighting using Mains-Borne Signalling with Photocell Control in a Drawing Office

The control technology used in the ECS 500 system installed at Portsmouth is fairly conventional. However, its use for lighting control remains relatively novel with presently only a small but rapidly increasing fraction of the total potential market having been penetrated. One of the reasons for this is the high proportion of total costs accounted for by installation, especially when retrofitted to existing buildings. This limits the detail of control that can be provided. It would be highly uneconomic, for example, to try to provide control of individual luminaires in this way in order to cater for a complete pattern of occupancy and working schedules, etc. The cost of wiring the central controller to individual fittings would be prohibitive.

If highly detailed patterns of lighting control are nevertheless required, more advanced technology is becoming available to enable this to be provided in a cost-effective way. In particular, it is now possible to exploit the existing mains wiring for transmission of the control signals, thus obviating the need for separate control links. This type of technology has widespread potential application for the remote control of electrical loads and is already in limited use in road lighting and electrical heating of local authority apartment blocks. Its use for lighting control in an individual building is particularly novel at the present time, and a monitored demonstration project has been set up to assess prototype equipment developed jointly by GEC Meters Ltd and Osram (GEC) and installed in a drawing office of GEC Turbine Generators Ltd at their Trafford Park works in Manchester.

THE ECS 500 AUTOMATIC LIGHTING CONTROL SYSTEM USED BY PORTSMOUTH CITY COUNCIL

Figure 3.3: The central time switch.

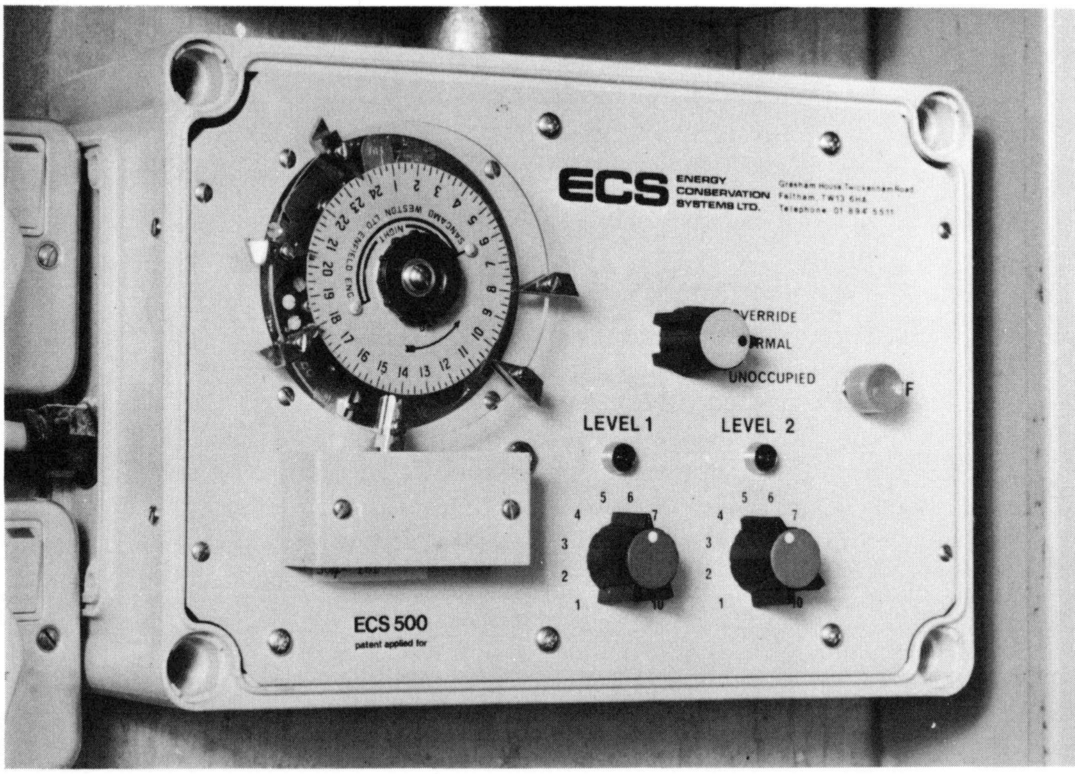

Source: ETSU

Figure 3.4: Manual override panel to enable lighting to be selectively restored after an automatic off-switching. The arrangement of buttons corresponds to the lighting layout on the particular floor.

Source: ETSU

Figure 3.5: Typical modulation of AC power waveform used in the 'Cyclocontrol' mains-borne signalling technique. The relative amplitude of the control signal is typically 5 per cent of the peak AC amplitude.

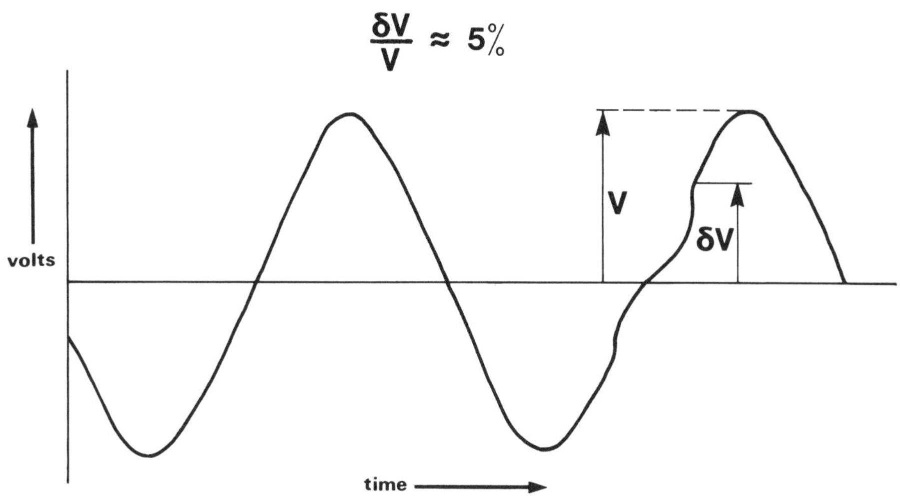

The lighting control system being demonstrated exploits an existing mains-borne signalling system developed by GEC Meters and called the 'Cyclocontrol'. This provides a slight modulation of the normal AC supply waveform for a fixed number of cycles such that it remains within the Electricity Board specification for voltage and frequency, etc but forms a recognisable message or address detectable by a special receiver/actuator unit. What happens is that thyristors in the control unit momentarily short circuit the mains supply a few millivolts either side of the zero voltage cross-over point for a fixed number of successive cycles. The effect is to alter the slope of the waveform at this point (Figure 3.5). This slight distortion will propagate throughout the mains circuitry of the building but will not pass through transformers. Because the waveform remains within specified limits, this control technique is unlikely to cause interference with other electronic equipment.

The Osram (GEC) contribution to the development is the special microprocessor-based low-cost receiver/actuator unit which is designed to be easily fitted to existing fluorescent luminaires. Each receiver can be set to one of four control categories, corresponding to up to four separate control programs stored in the memory of the central control unit. These programs can control according to time and/or type of day or control perimeter lighting using photocells. Not only can lights be switched on and off, but they can also be set to intermediate levels of output (40 per cent and 70 per cent of maximum) by means of a separate dimming circuit in the receiver unit.

The receiver units are activated by an address signal (corresponding to the selected control strategy) from the central control unit. The dimming unit will then respond to subsequent control messages activated by the time programs or the photocell daylight readings and the transition between light output levels is achieved by an apparently smooth series of small step changes.

This project is also being independently monitored for the Department of Energy. Savings to date have not been dramatic (up to eight per cent of previous lighting consumption), largely because the central programming and operation of the receiver units have still to be adjusted and improved. However, the monitoring report (which is now available from ETSU) expresses confidence that this prototype unit will eventually develop into a worthwhile lighting control system capable of achieving savings of up to 40 per cent of previous consumption at Trafford Park. No electrical interference with other electronic equipment has been reported. Furthermore, it is interesting to note that the difference in apparent light output between the 70 per cent and 100 per cent levels is almost imperceptible, even though substantial savings are being made. The transition between lighting levels appears particularly smooth.

The principal advantage of a mains-borne signalling system of this type is its potential ability to provide flexible control of lighting at minimum installation cost. For example, lighting can be programmed in different ways on a single floor to cater for different patterns of work. Draughtsmen may work different hours from (say) clerical staff occupying the same area; providing a separate lighting control program for the respective overhead luminaires (using the simple technology described in the previous example at Portsmouth) would have been far too expensive. On the other hand, the greater inherent complexity of the mains-borne equivalent system carries its own potential drawbacks; there have been a number of teething problems with the GEC system at Trafford Park and it is not a simple matter for the user to tune the system to his actual requirements. A separate behavioural study has shown that levels of satisfaction and acceptance among the drawing office staff have been rather low so far. Nevertheless, the technique has considerable future potential.

Time-Based Off-Switching with Local Reset of Individual Luminaires in an Office Building

Both examples described above include bulk switching of lighting on and off according to time/type of day and ambient daylight levels. However, as we have previously observed, people are very good at switching lights on; it is switching them off again that is often neglected. Consequently, there is much to be gained from a control system that concentrates on turning lights off where appropriate, while providing comprehensive manual facilities for individual lights to be turned back on again, if required. If the lights are switched off at strategic times (such as mid-morning and afternoon breaks, lunchtimes and normal leaving times) the pattern of manual reset switching during the intervening periods is likely to correspond reasonably closely to the actual lighting requirements.

A demonstration project has been supported by the Department of Energy to study the performance and cost-effectiveness of a simple system of the above type in a typical multi-storey office building. The project concerns an ECS 450 system, again by ECS Energy Conservation Systems Ltd, installed in the 8,600 sq metre civic offices of Bradford Metropolitan City Council. This building (Figure 3.6) is a fairly conventional office block with seven floors.

The lighting control system comprises a microprocessor-based central time programmer (Figure 3.7) which provides short power interrupts to the lighting distribution circuits. These power interrupts cause the special pull-cord reset switches installed on each luminaire (Figure 3.8) to latch in the 'off' position. Occupants can restore individual fittings to 'on' by pulling the cord. The central programmer can provide up to eight separate interrupt schedules for different circuits to cater for the varying requirements of the building. On smaller scale installations, the power interrupts could be provided by the conventional wall switches and Bradford City Council embarked on an initial pilot scale project in this way.

An interim report is now available for this project and Figure 3.9 illustrates the principal findings to date. It can be seen that substantial savings of about 42 per cent of previous lighting electricity consumption are being achieved. This represents a payback time of under two years on the original investment of £13,000. Furthermore, a behavioural study has shown very high levels of satisfaction and acceptance among the Council staff.

This system thus provides a very effective solution to the problems of matching lighting supply and demand in open plan office spaces and its simplicity of operation is regarded as a distinct advantage. However, it might not necessarily be appropriate for all types of spaces such as cellular offices; prospective users need to carefully assess the applicability of such a system to their own individual circumstances, bearing in mind the way in which their building is operated. A particular drawback is the aesthetics of ceiling mounted pull-cords, which might not be acceptable in a prestige building.

THE BRADFORD METROPOLITAN CITY COUNCIL AUTOMATIC LIGHTING CONTROL PROJECT

Figure 3.6: The civic offices—a fairly conventional speculatively built office block.

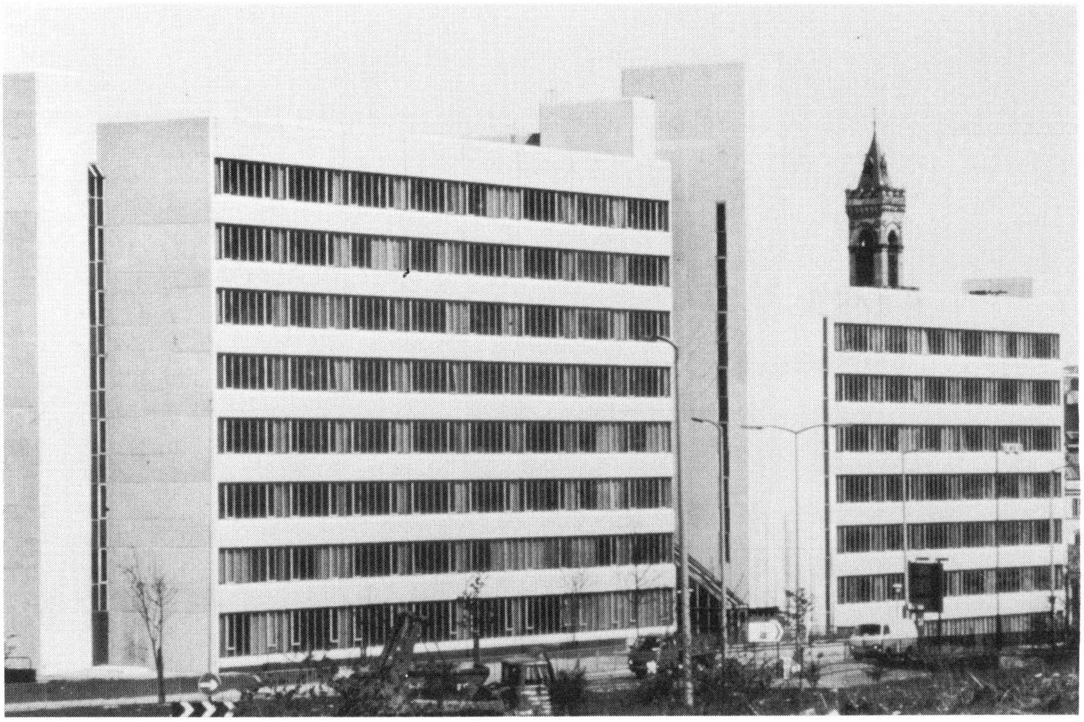

Source: ECS—Energy Conservation Systems Ltd

Figure 3.7: The ECS microprocessor-based time programmer used to provide the lighting power interrupts.

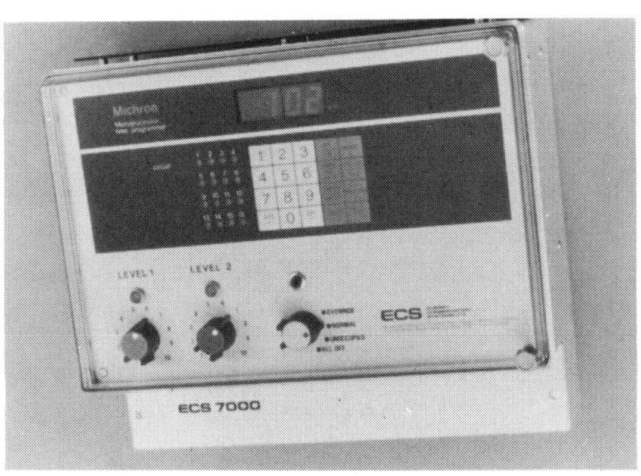

Source: ECS—Energy Conservation Systems Ltd

Figure 3.8: The ECS pull-cord lighting reset switch.

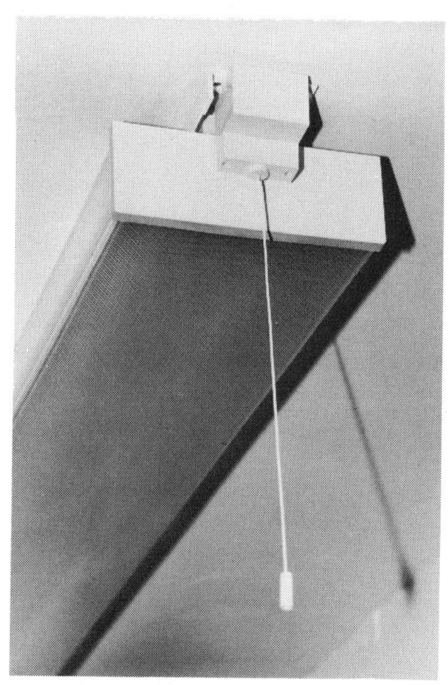

Source: ECS—Energy Conservation Systems Ltd

THE BRADFORD METROPOLITAN CITY COUNCIL AUTOMATIC LIGHTING CONTROL PROJECT

Figure 3.9: Mean day and night lighting electricity consumptions for the south side of the fifth floor of the civic office building. This shows the drop in lighting consumption due to installation of the reset switches.

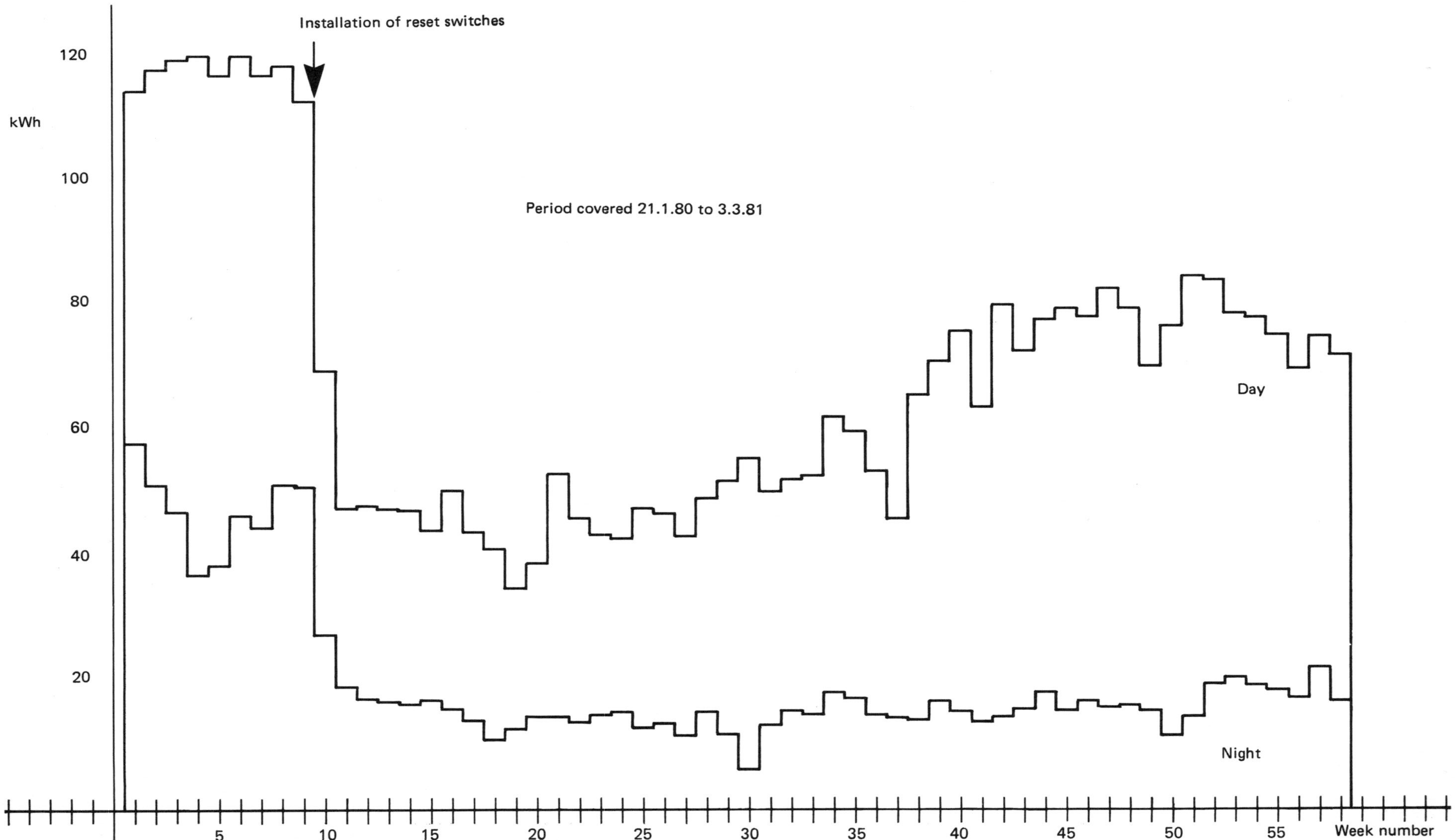

Centralised Mains Borne Off-Switching of Lights in a Refurbished Office Building with Local Reset via Infra-Red Controllers

This project has been supported as a demonstration in the headquarters building of Lloyds Bank in Cannon Street, London. It involves equipment recently developed by Home Automation of Harpenden to meet the particular requirements of the bank; it provides cost-effective automatic off-switching of lighting with minimum installation costs and visual intrusion and maximum flexibility of programming. This will help the bank to cater for possible substantial future changes in the way the building is used.

Home Automation adapted their 'Ripul' mains borne signalling system (which has a wide range of other potential uses, including in the domestic sector) to address about 500 local slave units (Figure 3.10) located around the building, with normally one per office where cellular spaces are concerned. The control signals are derived from high frequency (125-140 kHz) pulses injected onto the mains supply. These are picked up by the slave units which are able to switch off up to four adjacent luminaires according to time/type of day and measured ambient daylight levels. Different categories of slave unit can receive different switching schedules according to programs stored in a Seaboard comprehensive energy management system interfaced with the Ripul signal generator.

In terms of the way it operates, the Home Automation system can be regarded as a more sophisticated equivalent of the ECS 450 system installed at Bradford. However, instead of using pull-cords for local manual reset, Home Automation provide hand-held infra-red controllers similar to the type used with domestic TV sets, or similar cordless wall-mounted units (Figure 3.11). Each controller has four push-buttons corresponding to the four luminaires controlled by the local slave unit. By aiming the controller at the infra-red receiver on a nearby slave unit, an occupant is able to switch local lighting on without necessarily having to get up from his desk. By operating in this way, the Home Automation system is potentially appropriate for a wide range of office environments, both cellular and open-plan. In particular, the fact that there is no physical connection between the signal generator, slave units and local controllers (other than existing mains wiring), means that installation costs are relatively low and system flexibility high.

The energy savings and overall performance of this demonstration are being monitored over a period of two years before and after full commissioning of the system. In addition, a separate behavioural study has been conducted on the project. A full monitoring report is now available from ETSU. This shows that savings of around 40 per cent of previous lighting consumption are being achieved. It is difficult to quantify the payback achieved because some form of light switching was required in the refurbishment of this building and the additional costs of installing the Home Automation system over a conventional wiring and switching arrangement were small. In existing buildings, where there will be a direct additional retrofit cost, paybacks should be around one to three years, depending on exact local circumstances.

No problems with electrical interference from the Ripul system have been encountered in this project. However, if these mains-borne techniques gain widespread acceptance, there is likely to be a need to avoid 'crosstalk' between neighbouring buildings equipped with similar systems; suitable filters will be required where no isolating transformers are fitted.

There have been some difficulties with ensuring that the signal strengths are adequate throughout the building, particularly in areas furthest from the controller (i.e. on the top floor). Also, adequate battery strength on the infra-red controllers and a clear line of sight to the slave units must be maintained.

The behavioural studies show that the system has been well received by the occupants and that good use is being made of it. Indeed, some people feel that comfort conditions have actually been improved because the building was previously overlit. However, some minor problems with the operation of the remote reset units have arisen from low batteries, limited range and misunderstandings about how the system operates. Routine maintenance procedures are being adjusted as a result and Home Automation have now enhanced the specification of the system in order to improve the ease of operation and sensitivity of the local reset facility. In addition, there are now many more channels on the central control unit.

Figure 3.10: Home Automation ceiling-mounted lighting control slave unit (to the left of the luminaire) incorporating receivers for signals from 'Ripul' mains-borne transmitter and infra-red reset switches.

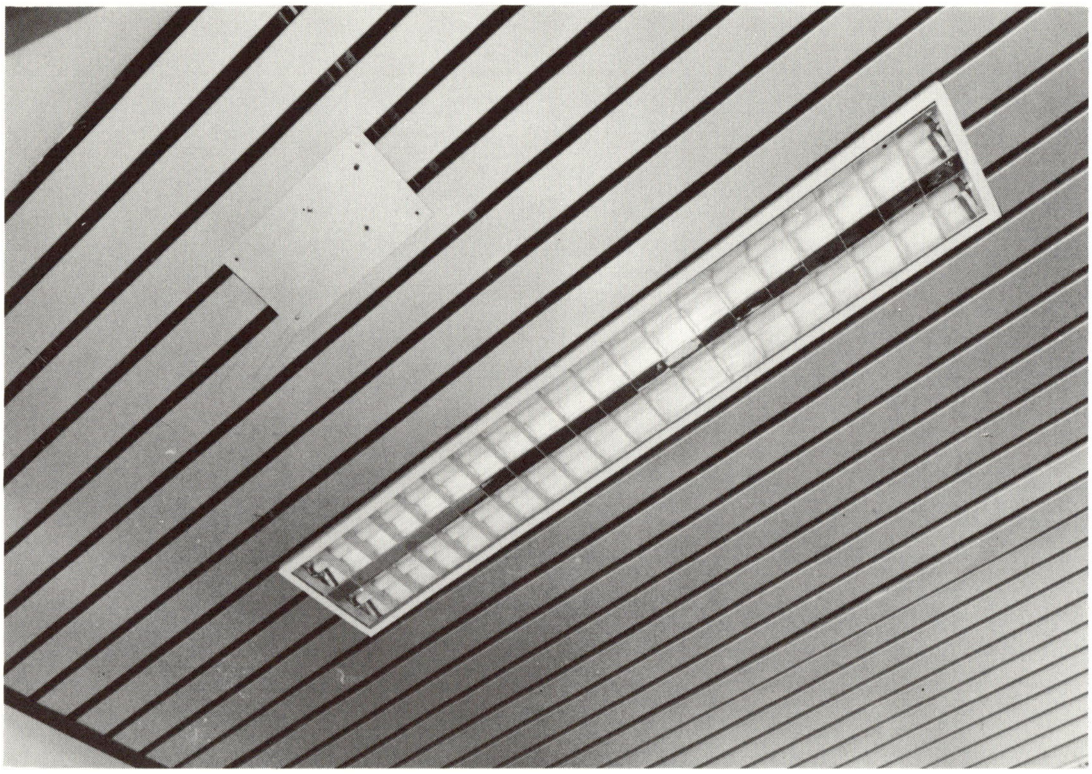

Source: ETSU

Figure 3.11: The wall-mounted infra-red light reset switch. There is no cable link between this unit and lighting circuits.

Source: ETSU

Time-Based Off-Switching of Arrays of Luminaires in a Cosmetics Packaging Plant with Local Manual Reset

All of the preceding examples of automatic lighting control concern office buildings. Other types of building have similar problems of unnecessary lighting usage, although these can arise from somewhat different circumstances. For example, in industrial buildings the need for lighting may vary according to the levels of production. Such a requirement for lighting control has arisen at the cosmetics packaging plant of Avon Cosmetics in Northampton. Natural daylight makes no significant contribution to lighting levels in the main packaging hall (Figure 3.12) and consequently all the lighting tended to be on for most of the time, regardless of how many of the several dozen packaging lines were in operation. Lights would be put on by security staff in early morning (0730) and remain on until about 2300 hours. Actual production hours are 0800-1630 and 1800-2200 hours and, even during these periods, certain filling lines would be out of use and not require illumination.

What Avon Cosmetics have done with Energy Conservation Demonstration Projects Scheme support is to re-arrange the power cabling for lighting in the packaging hall and to connect in a system of relays linked to an electromechanical time switch. This enables varying proportions of the lighting (i.e. 10, 30, 60 and 90 per cent) to be extinguished at strategic times during the working day (e.g. meal breaks, etc). Following an automatic switching, the lighting will stay at the levels set until the next such switching or until individual rows are reset to full 'on' by means of a local conveniently located switching panel (Figure 3.13). Thus individual packaging lines can be illuminated as required.

The complete installation was designed and supplied by Floorplan Electrical Ltd (Switchplan Division) of Wolverhampton and this demonstration project is being independently monitored over a one-year period of usage. Lighting consumption will be compared with historical sub-metered data previously collected by Avon. Analysis to date suggests that savings of about 23 per cent on the previous consumption of about 14,000 kWh per week are being achieved, giving a payback of just over two years on the original capital/installation costs of £16,700.

Automatic On/Off-Switching of Lighting in School Classrooms on the Basis of Occupancy

All of the lighting control strategies described above rely mainly on time-based switching to achieve a reasonable match between lighting needs and supply. A relationship with local occupancy (e.g. in a cellular office) is achieved by relying on a manual reset to 'on' following an automatic 'off'. These strategies work well in environments that are in regular use by a single set of occupants (e.g. in office spaces). However, where spaces are subject to highly intermittent use by a number of different groups (e.g. in school classrooms), a different control strategy based on actual occupancy should be more appropriate.

A project along these lines is presently being implemented by Wolverhampton Metropolitan Borough Council. It involves the use of special acoustic switches (Figure 3.14) developed by Allen Martin Electronics Ltd, also of Wolverhampton. These replace conventional wall switches and contain a microphone and photoelectric sensors (Figure 3.15); lights are switched off if no significant sound is heard for a preset time, which is adjustable between a few seconds and several minutes. If noise above an adjustable threshold level is subsequently detected, lights will be turned on again, providing the measured ambient lighting is below a preset level. In this mode, the switch also acts as a security device; however, this on-switching facility can be cancelled by means of a simple internal disconnection if it is not required.

Wolverhampton MBC are installing over 100 of these devices on a trial basis in a large technical college (Bilston) and a small primary school (Elston Hall). The savings are not expected to be as high as in the office buildings projects because the normally intermittent use of school spaces tends to ensure regular re-assessment of lighting needs and so wastage is often at a lower level. Actual savings are being monitored over a 12-month period of use but are expected to be around 15 per cent of previous consumption; this gives a likely payback period of about three years on a total investment of £10,000.

THE AVON COSMETICS AUTOMATIC LIGHTING CONTROL PROJECT

Figure 3.12: The main packaging hall. Prior to installation of the Switchplan lighting control system, most of the artificial lighting remained in use regardless of production requirements.

Source: ETSU

Figure 3.13: A local reset switch panel enabling lighting to be restored over a particular production line if it has been automatically switched off.

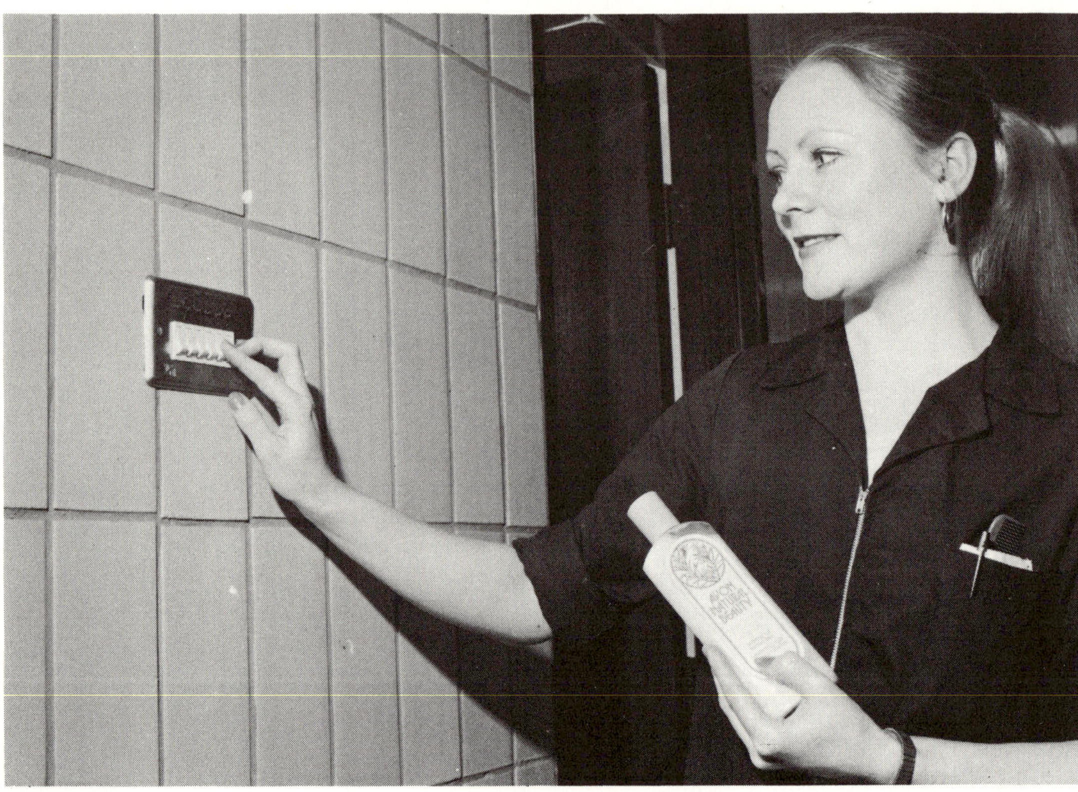

Source: ETSU

Figure 3.14: The Allen-Martin acoustic/photoelectric light switch. This replaces the normal light switch.

Source: Allen-Martin Electronics Ltd

Figure 3.15: Schematic diagram of the circuitry of the Allen-Martin light switch.

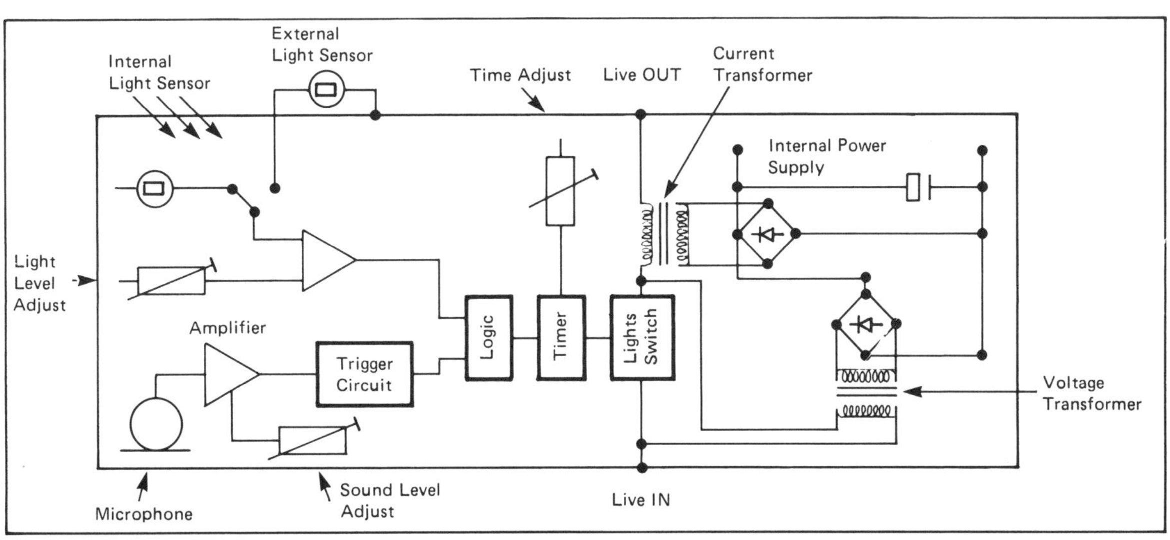

Source: Allen-Martin Electronics Ltd

It can be appreciated from the above examples that identifying the opportunities for a cost-effective investment in lighting control is not necessarily straightforward. Nor is selecting appropriate equipment and making full and proper use of it. These problems will be discussed in detail in subsequent chapters. However, a useful decision chart (Figure 3.16) has been developed by the Building Research Establishment; it is intended to help prospective users of automatic lighting control systems to clarify their initial ideas.

Boiler Control

Improved control of combustion has particular importance for national energy savings because the great bulk of primary fuel (oil, coal and gas) is initially burned in boilers and furnaces of all types, for example in power stations or steam raising plant. Modest improvements in overall boiler efficiency can give rise to very substantial national savings, once they are adopted widely. A particularly large market is the boilers used for space heating and hot water production in buildings.

There are three main ways in which improved controls can increase combustion efficiency, namely:

—oxygen trim (i.e. fuel/air mixture control)

—boiler sequencing (for a multi-boiler installation)

—boilerhouse management (for a large and complex installation).

Oxygen Trim

In boilers, as in all other fossil fuel combustion systems like internal combustion engines, it is important but very difficult to ensure that the correct ratio of fuel to air is obtained under all firing conditions. If there is excess air, some of the heat of combustion will be consumed in heating this air up and so will be lost up the chimney. Alternatively, if too little air is supplied, combustion will be incomplete and problems of smoke or soot might arise.

Clearly, therefore, there is an optimum ratio that varies according to different firing conditions. Ideally, this ratio should correspond to the stoichiometric ratio, which is the molecular proportions of air and fuel in the chemical equation of combustion. However, this only applies under conditions of perfect fuel/air mixing; because the mixing is always imperfect, there will be a need for a slight but varying excess of air to compensate.

On very old boilers (for example of the so-called Lancashire type), the excess oxygen levels would often be very substantial (i.e. 10-50 per cent) and largely uncontrolled. This, combined with other design factors such as inadequate insulation and poor heat transfer between flue gases and the contained water would commonly result in overall combustion efficiencies (ratio of useful heat exported to heat raised) as low as 60 per cent.

More modern boilers will conventionally have the fuel/air mixture ratio controlled by an adjustable mechanical linkage between the main fuel valve and air dampers to maintain the correct values over the complete firing range. Better overall physical design coupled with good standards of operation and maintenance by boilerhouse staff can routinely achieve efficiencies around 80 per cent. Unfortunately, these conventional control systems are prone to wear and maladjustment and cannot react quickly to changes in the operating condition of the boiler (e.g. to compensate for burner wear).

Modern oxygen trim systems are able to provide continuous interactive monitoring and control so that boiler combustion efficiencies can be maintained at significantly higher sustained overall levels over long operating periods; typical improvements of one to three per cent can be obtained over the best conventional practices. However, as we shall see, high technology is no substitute for good standards of operation and maintenance. Furthermore, the likely achievable savings are usually difficult to see against the normal background of varying fuel consumption.

Figure 3.16: Decision chart developed by BRE to assist in the selection of appropriate lighting control strategies. (Reproduced from Building Research Establishment Digest 272, HMSO Crown copyright).

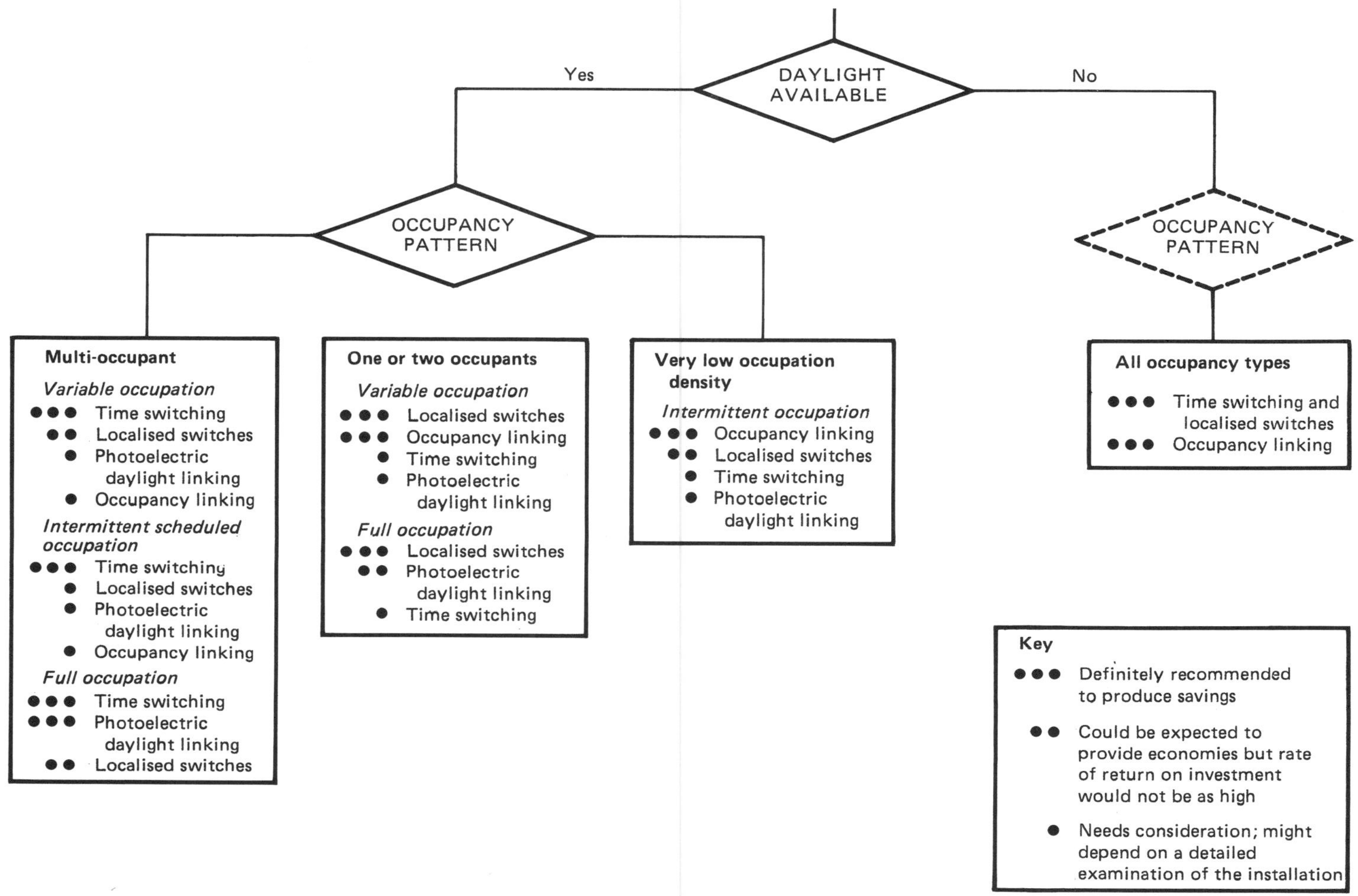

There are several techniques presently available for implementing fuel/air mixture optimisation. However, they all rely on the following main components:

–a sensor, to monitor continuously temperatures and concentrations of oxygen, carbon monoxide and/or dioxide (according to sensor type) in the flue gases; these parameters form the basis of the optimisation procedures

–an intelligent device (mini-computer or microprocessor-based) to receive and process the sensor signals and to determine control actions according to the operating software contained in its memories

–appropriate actuators to adjust air dampers, etc according to control instructions sent out from the main processing unit.

The likely fuel savings achievable through oxygen trim techniques may be small in percentage terms but will still be worthwhile. However, the magnitude of the likely savings place a strict upper limit on the amount of money that can be spent on such a system if it is going to represent a reasonable investment opportunity; the various techniques available are generally in different cost ranges so the choice of system type is determined mainly by the size of boiler.

At the lower end of the boiler size range (below about 80,000 lb steam per hour output) the limitations imposed by cost-effectiveness restrict combustion control to simple systems sensing excess oxygen levels in flue gas. This is normally achieved by means of a zirconia Nernst cell consisting of a re-entrant zirconia tube coated on its inner and outer surfaces by thin conducting layers. The inner cavity is open to atmosphere which forms the reference oxygen concentration (21 per cent by volume). A heating element raises the temperature of the zirconia to around 700°C, i.e. well above typical flue gas temperatures of around 450°C. At this temperature, the zirconia acts as a semiconductor which, under the influence of different oxygen concentrations, generates a voltage difference between its inner and outer surfaces. This electric potential forms the sensor signal and is converted to a flue gas excess oxygen reading by software in the central control unit. This then transmits the appropriate control signals to the air dampers. Figure 3.17 illustrates a typical oxygen trim control network, namely for the ETC Boilermizer system to be described later.

In addition to providing closed loop control of fuel/air mixture ratios, modern oxygen trim systems of this type can also display and record information on important boiler parameters such as flue gas temperature and firing rate and can provide alarm messages when these parameters go outside preset limits. Furthermore, by also measuring the temperature of the incoming air, the computing capabilities of the microprocessor can be exploited to provide a calculated reading of instantaneous combustion efficiency.

The principal advantage of the zirconia cell oxygen trim system is its relatively low cost, which has been reduced from a minimum of around £10,000 to nearer £2,000 in recent years by exploitation of cheaper microprocessor technology and sensors. This has enabled the cost-effectiveness of the technique to be extended down to boilers with outputs of about 5,000 lb steam per hour.

However, the technology does have several potential drawbacks:

–the control signal is vulnerable to errors caused by air leaks upstream of the sensor position

–the requirement to heat the probe to 700°C introduces the potential danger of explosion of unburnt fuel (for example, on boiler startup)

–the zirconia probes are a relatively expensive proportion (30-40 per cent) of the total system costs. Furthermore, they are somewhat fragile, prone to mishandling and could require replacement after a few years' use. This introduces a significant maintenance cost

–although suitable for oil- and gas-fired plant, the difficulties of monitoring and controlling solid fuel make zirconia-based oxygen trim systems unsuitable at present for this application. However, development work now underway may resolve this problem.

Figure 3.17: Schematic diagram of the operation of the ETC 'Boilermizer' oxygen trim system.

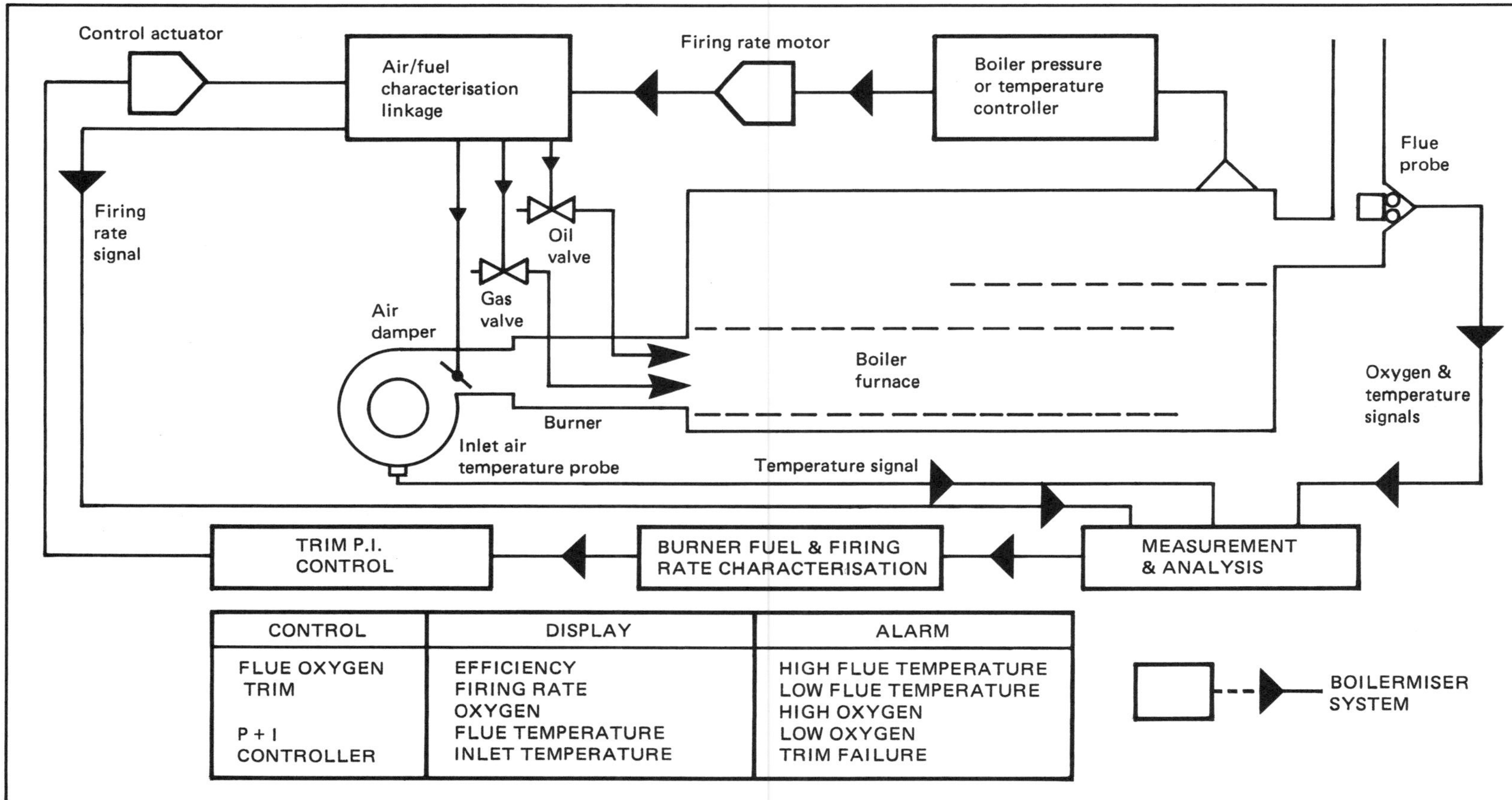

Source: Energy Technology & Control Ltd

So far, oxygen trim systems have not become standard equipment on new package boilers and largely remain a retrofit opportunity for existing boilers. Examples of two such applications of low-cost oxygen trim systems can be found within the Energy Conservation Demonstration Projects Scheme. Both involve a system developed by Energy Technology and Control of Lewes, Sussex called the 'Boilermizer'. The first project utilises a prototype version on a 16,000 lb steam per hour oil-fired boiler at the Spillers petfood cannery in Wisbech (Figures 3.18-3.20). The boiler is one of three in a boilerhouse providing steam to a range of pressure cookers. Whilst this is not strictly a 'buildings' applications, the boilerhouse is of a size typically found supplying steam for space and hot water heating on medium to large sites.

At Spillers, variations in production output can cause steam demand from individual boilers to fluctuate by factors of up to ten during a typical week, reflecting the nature of batch processing. Consequently, boilers frequently operated over their full firing range and it was not easy to maintain peak combustion efficiency. Standards of boilerhouse operation and maintenance were generally high and efficiencies of around 80 per cent were commonly achieved. Nevertheless, it was felt that scope for further improvements existed through the use of oxygen trim.

The performance and cost-effectiveness of the Boilermizer system at Spillers has been independently monitored and the final report is now available from ETSU. There are three principal findings.

Firstly the actual fuel savings were exceedingly difficult to identify and quantify, because the effect of installing the Boilermizer was small in comparison with the total effect of all other factors influencing boiler efficiency (the condition of the burners, atmospheric pressure, humidity and temperature, fouling of heat exchanger surfaces, etc). The low signal-to-noise ratio posed a severe problem requiring very careful statistical analysis to separate out the effect being studied.

Nevertheless, it is believed that an average overall saving of between 1.0-1.5 per cent of previous fuel consumption was being achieved following the installation of oxygen trim control (Figure 3.21). This represents a saving of £3,360, giving a payback of about 14 months on the original equipment cost of £4,500 (including £500 for a chart recording facility).

However, it must be remembered that these savings were only established by a specialist organisation carrying out detailed monitoring at a cost considerably greater than that of the equipment itself. It would have been virtually impossible for the user organisation itself to have accurately quantified the actual benefits without committing a similar level of expenditure. This is unrealistic, so future users will need to take the savings achieved largely on trust.

Finally it was also confirmed that the savings achieved were the minimum likely to be obtained because of the especially good standards of housekeeping, operations and maintenance in the Spillers' boilerhouse. However, the extent to which oxygen trim systems can compensate for deficiencies in these areas is strictly limited and prospective users should definitely not regard these systems as substitutes for good conventional practices; rather, they should be looked upon as supplements or aids to such practices. Indeed, installation of oxygen trim on a badly set up and maintained boiler could well be a waste of money.

A second demonstration of the ETC Boilermizer system is taking place in the gas-fired boilerhouse of the University of Sussex near Brighton. This provides high pressure hot water for space and domestic water heating. The pattern of heat demand from this boilerhouse is quite different from that at Spillers and information arising from this project will add significantly to our understanding of the actual performance and benefits of this technology. At the time of writing, the final report had not been published, but it should be available by the end of 1984.

In the meantime, ETC have succeeded in selling a substantial number of similar devices to other users (100 plus by mid 1983), helped in part by the additional credibility obtained through the ECDPS reports. Furthermore, comparable low-cost equipment is now available from other firms, such as Westinghouse with their 'Minitrim' system. However, penetration to date remains a small fraction of the total market which comprises many thousands of industrial/commercial boilers in the small-to-medium size sector for which this type of equipment is intended. It will therefore be some time, before the full national potential for energy savings is realised.

Larger boilers can justify the costs of more sophisticated combustion control equipment. For example, infra-red spectroscopic analysis of the flue gases enables levels of carbon dioxide and

THE SPILLERS FOODS BOILER CONTROL PROJECT

Figure 3.18: The 16,000 lb steam per hour package boilers providing process steam for batch cooking at the Spillers petfood cannery in Wisbech, Cambs.

Source: ETSU

Figure 3.19: The ETC Boilermizer control unit.

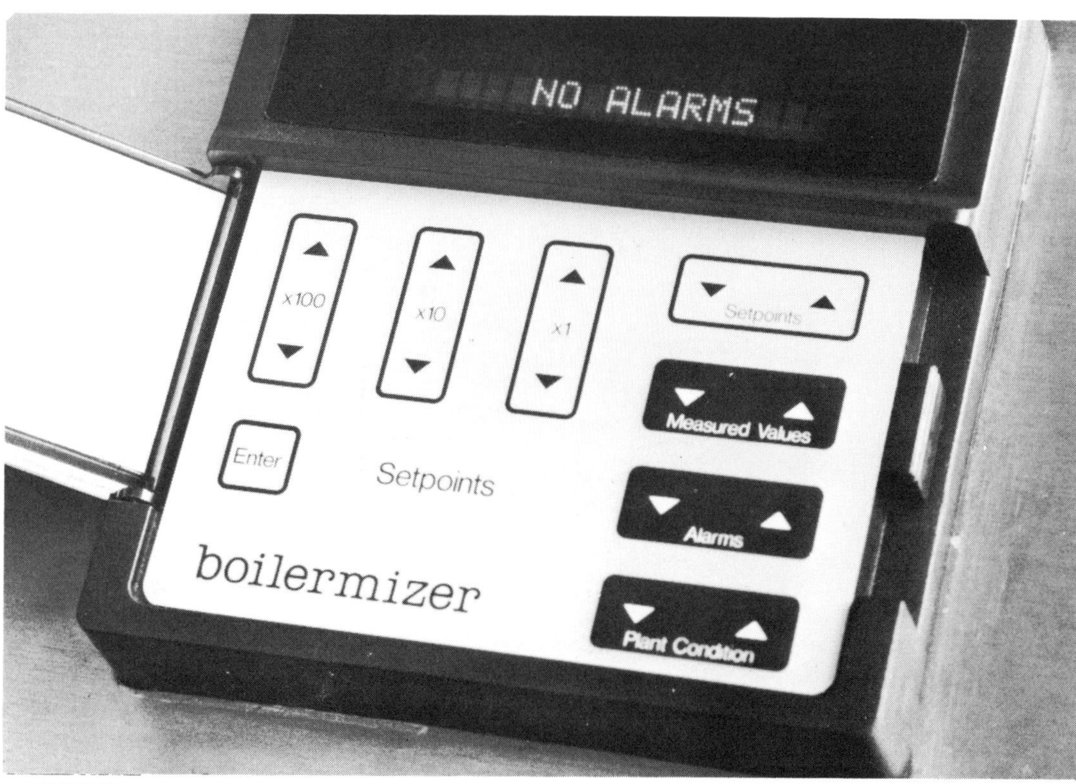

Source: ETSU

Figure 3.20: The ETC Boilermizer air damper actuator mechanism.

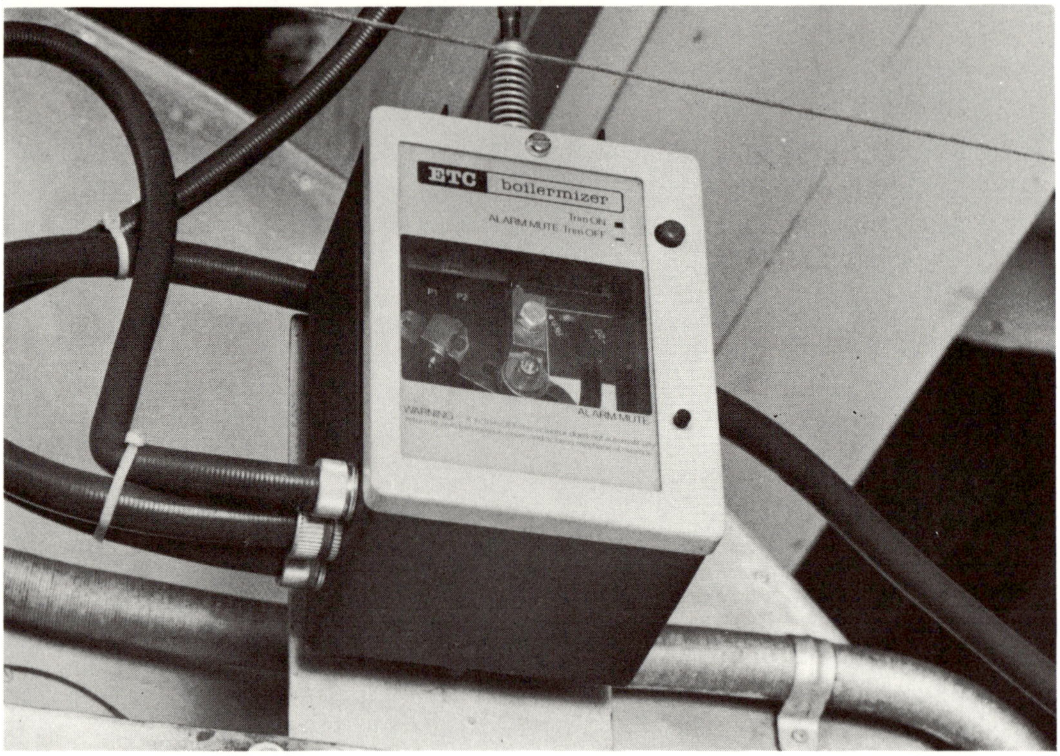

Source: ETSU

Figure 3.21: The effect of installing oxygen trim on excess oxygen and combustion efficiency of boiler operations. The tighter overall control represents about 1-1.5 per cent improvement in efficiency.

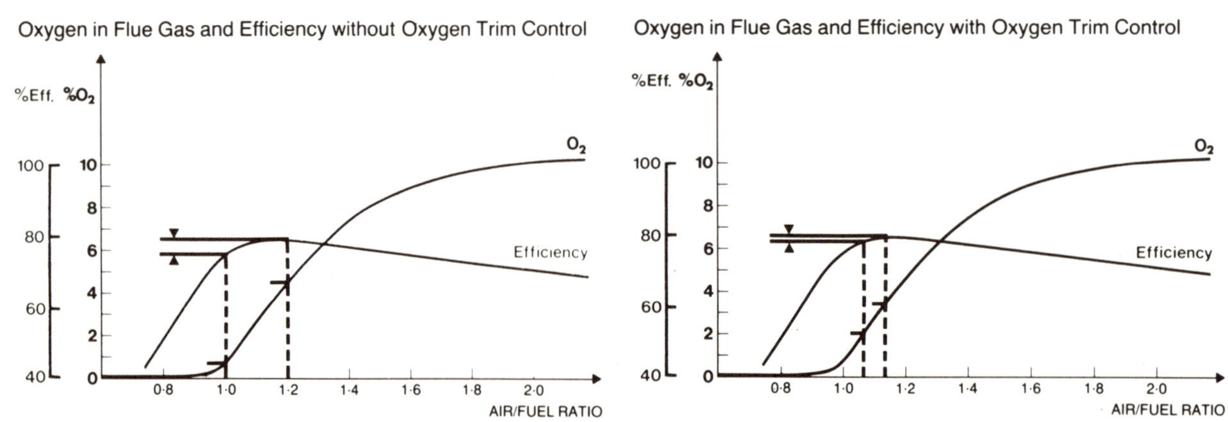

monoxide to be measured and used for control purposes. This technique carries the advantages of being non-intrusive into the flue gases and being able to scan across the entire width of the flue with a traversing light beam. Moreover, the readings are unaffected by air leaks. However, current costs of equipment of this type mean that it is likely to be economic only on very large boilers in the industrial processing or power generation sectors and it is not appropriate at present for typical applications in the buildings sector. An example of the technique used in combination with other boilerhouse management techniques will be briefly described later in this chapter.

Boiler Sequencing

In common with many other types of building services plant, boilers are often considerably oversized in order to provide a substantial margin of capacity. Under these circumstances, plant is liable to be operating at well below full load and, consequently, at less than peak efficiency. This problem can be exacerbated when there is a multiple installation with several boilers in a boilerhouse, all of which may just be ticking over at low efficiency.

Low-cost microprocessor-based equipment is now available to monitor heat demand and automatically schedule the operation of multiple boiler installations. This ensures that the minimum number are firing at any one time and that additional boilers are not brought on line until demand from the most lightly loaded boiler approaches its full output. Units of this type (Figure 3.22 shows a JEL unit) can be utilised either as a dedicated controller or as a component of a comprehensive energy management system.

Boilerhouse Management Systems

When boilerhouse installations become particularly large or the patterns of steam demand highly complex, there may be considerable potential for savings through centralised monitoring and control of many boiler parameters; these include oxygen trim and boiler sequencing, as well as matching steam supply with demand. Few systems of this type are in operation to date, and most of them are to be found on very large boiler installations such as CEGB power stations. Present costs preclude the use of boilerhouse management systems on most buildings applications where there is no process heat demand; however, there are several installations on large sites such as universities. There is a good example at the University of Liverpool.

A demonstration project involving a particularly sophisticated example of this technology incorporates not only the principles of centralised boilerhouse management but also infra-red spectroscopic flue gas analysis as described earlier in this chapter. It is located at the Kemsley mills of the Bowater Paper Co, near Sittingbourne in Kent. The large boilerhouse contains six coal-fired boilers, with capacities ranging up to 450,000 lb steam per hour. They are used both for process heat and for electricity generation.

Figure 3.23 depicts the sophisticated central station; this carries out automatic optimisation and displays all major boiler parameters. In addition to minimising excess air heat losses, the system ensures that the boilers are controlled in response to demand so that the most efficient are meeting most of the load. It also maximises the output from the most efficient electricity generating set.

This project has been closely monitored over several years' operation. Unfortunately, no definite savings of coal were detected. This was due both to the severe technical difficulties of monitoring a solid fuelled system and to the relatively small savings expected (about three per cent); these were totally obscured by all the other influences on coal consumption.

Nevertheless, it is believed that such savings are possible and that further costs savings may be achieved through reductions in boilerhouse manning levels. The latter can be difficult to achieve without the wholehearted cooperation of boilerhouse staff; this may not be forthcoming if job losses are in the offing. However, even if manpower savings are forthcoming, it is unlikely that an installation of this type can be justified for space and domestic hot water heating applications until equipment prices are significantly reduced.

Figure 3.22: The JEL boiler sequencing unit.

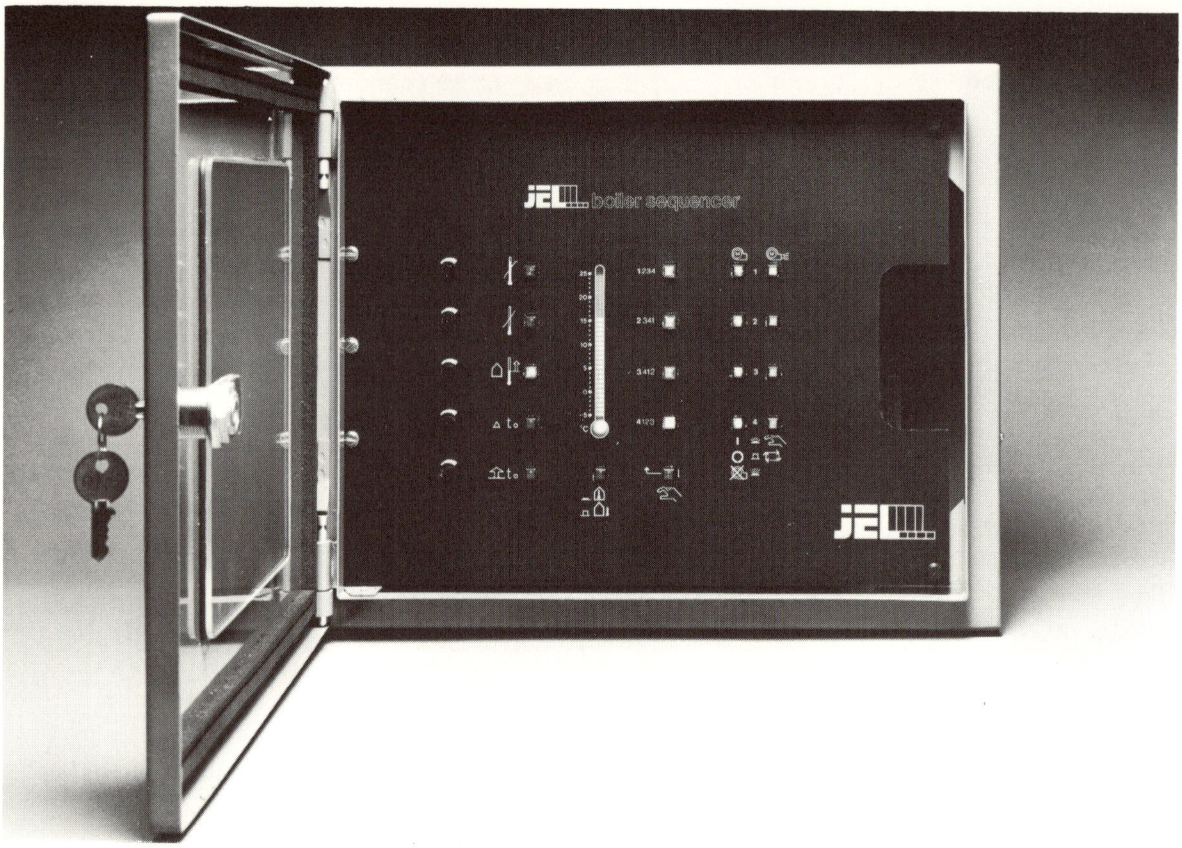

Source: JEL Energy Conservation Services Ltd

Figure 3.23: The central control station on a Measurex boilerhouse management system.

Source: ETSU

Heating Control

For many years, it has been standard practice to provide some form of automatic control of space heating in non-domestic buildings. However, this has usually been based on relatively simple electromechanical or hard-wired electronic systems using time-switches, bimetallic thermostats, electromechanical relays, and similar devices. Typical control functions would be:

–time-based on/off switching of hot water circulation pumps, perhaps with some simple adjustment for external air temperatures of the start time (i.e. with a so-called optimiser unit)

–modulation of the circulating water temperature according to external temperature, using three-way mixing valves (i.e. with a compensator unit)

–space temperature set point control, using the familiar room thermostats.

These technologies are very well established; they are capable of achieving substantial savings over purely manual methods of control, standards of which often degenerate; this results in excessive or inadequate comfort conditions and/or unnecessary supply of heat.

However, modern microprocessor technology is now able to offer considerable improvements in performance over these conventional systems. This modern technology is used in a large range of dedicated heating controllers introduced in recent years and in comprehensive energy management systems. Most building services engineers are already familiar with or are aware of these dedicated devices: they will not, therefore, be described in detail here, but the principal performance improvements that have stemmed from the application of modern microelectronics will be summarised.

Time programming can be much more detailed and flexible than with conventional time-switches, which might only be capable of providing perhaps one 'on' and 'off' for each day of the week or at most several on/off switchings over a 24 hour cycle. Microprocessor-based systems can provide many separate programs catering for perhaps several 'ons' and 'offs' for each of the days of the week (or year even) and possibly covering more than one channel. This enables the time variations in the heating requirements of a building to be met much more exactly.

The detailed programming capability provided by the microprocessor can be extended to allow several different zones of a building to be controlled separately with their own time programs and temperature settings. This can be of substantial value in most non-domestic buildings, where patterns of occupancy and mode of use can vary substantially in different areas. For example, the heating needs of sedentary 'nine till five' office workers are unlikely to correspond to those of industrial shift workers.

The latest microprocessor-based systems employ solid-state temperature sensors (e.g. thermistors). These are much more accurate and allow better spacial coverage of temperature measurement, so providing more precise control than the old bi-metallic thermostats. The latter devices are normally restricted to one per zone (or building) and suffer a substantial temperature dead-band (typically several degrees) between upper (off-switching) and lower (on-switching) thresholds. This can give rise to significant excess heating consumption because a $1^{o}C$ excess space temperature causes a 5-20 per cent increase in the difference between internal and external temperature and a corresponding increase in heat loss through the building fabric. Furthermore, several solid-state sensors can be employed to determine space temperatures in different areas of a building or zone and provide, through averaging, a much more representative measurement than is possible with a single room thermostat.

The power of the microprocessor can be further exploited to provide a range of additional convenient features such as a 'holiday skip', whereby heating can be disabled for a fixed period (usually up to 99 days) followed by resumption of the normal programs; this is a great aid to avoidance of inadvertent heating during shutdown periods. If heat is required during such a period, a manual override for a fixed duration (perhaps several hours) is normally provided. A frost protection facility will cause the heating to come on if internal temperatures fall to around $5^{o}C$. Furthermore, the effects of regular events such as the commencement of daylight saving time or leap years can often be taken into account automatically.

It is often inconvenient and 'fiddly' to alter the programs on conventional heating controllers and information output is usually minimal. Modern microprocessor-based dedicated devices have calculator-style keyboards for easy entry of program parameters; these can be presented on a digital display together with important information on space temperature readings. Further facilities can be provided with the comprehensive energy management systems to be described in the next chapter.

Units with most of the features described above can now be obtained for prices approaching those applicable to the domestic sector, i.e. down to around £50.00. Figure 3.24 illustrates a typical low-cost unit. The programs for these cheaper units will have fixed start/stop times, but a few hundred pounds more (plus installation costs) will provide the capability to optimise these times against the external temperature and thermal inertia of the building (or individual zones). The more sophisticated conventional devices do this for start times using 'hard-wired' fixed programs set up in the factory, with limited manual adjustment on the unit to take actual local circumstances into account; microprocessor-based units achieve this automatically, using self-adaptive software that 'learns' the thermal behaviour of the building by comparing actual temperatures with the target at a given time and making appropriate adjustments so that the difference is gradually reduced. The same can also be done for stop times, so that in mild weather the start of heating is automatically delayed and the stop advanced; the opposite will happen in cold weather. This feature, combined with accurate space temperature control and separate zone programming, enables the required comfort conditions to be provided at minimum cost. Figure 3.25 shows a typical unit with these more advanced control features.

The fuel savings compared to conventional control practice will depend on previous circumstances and can even be negative if comfort conditions were inadequate before. However, in practice this is rarely the case and actual savings can be expected to be in the range 10-20 per cent.

Electrical Load Management

This is an important area for improved monitoring and control, not least because of the substantial proportion of fuel costs nowadays accounted for by electricity in buildings with air-conditioning and high lighting loads. The problem is that the typical electricity bill for non-domestic consumers contains two components; one relates to the units consumed and the other to the maximum electrical demand in any fixed period (normally each half-hour). Substantial cost penalties can be incurred if the demand exceeds pre-determined limits.

Microprocessor technology can now be exploited to continuously monitor electrical demand and, if this begins to approach the maximum demand limitations (say above 80 per cent), selected loads can be shed according to pre-set priorities. Clearly, this technique can only be exploited if a building contains loads that can be shed safely without causing unacceptable damage or inconvenience. Water heaters, refrigeration plant, and similar systems where there is considerable energy inertia, can normally be set to high shedding priority without difficulty. The length of time the equipment is off is unlikely to be prolonged, because peak demand often occurs over quite short periods of an hour or so (e.g. in winter afternoons). It is normally inappropriate to shed lighting or production plant, because the consequent dangers, irritation and financial losses may not justify the saving in electricity costs.

Load shedding is principally a cost-avoidance rather than an energy conservation opportunity. However, both types of saving can be achieved through the second electrical load management technique of load cycling. This can be used where there is a multiplicity of items of equipment, all operating in a similar way. An example would be ventilation fans in (say) a factory; if one of these fans is switched off for a short period at any one time, the effects on comfort conditions are unlikely to be noticed but the benefits in reduction of electricity consumption can be significant.

American manufacturers have been in the forefront of development of suitable equipment for electrical load management. This reflects the dominance of electricity costs in their buildings,

Figure 3.24: A typical low-cost heating controller.

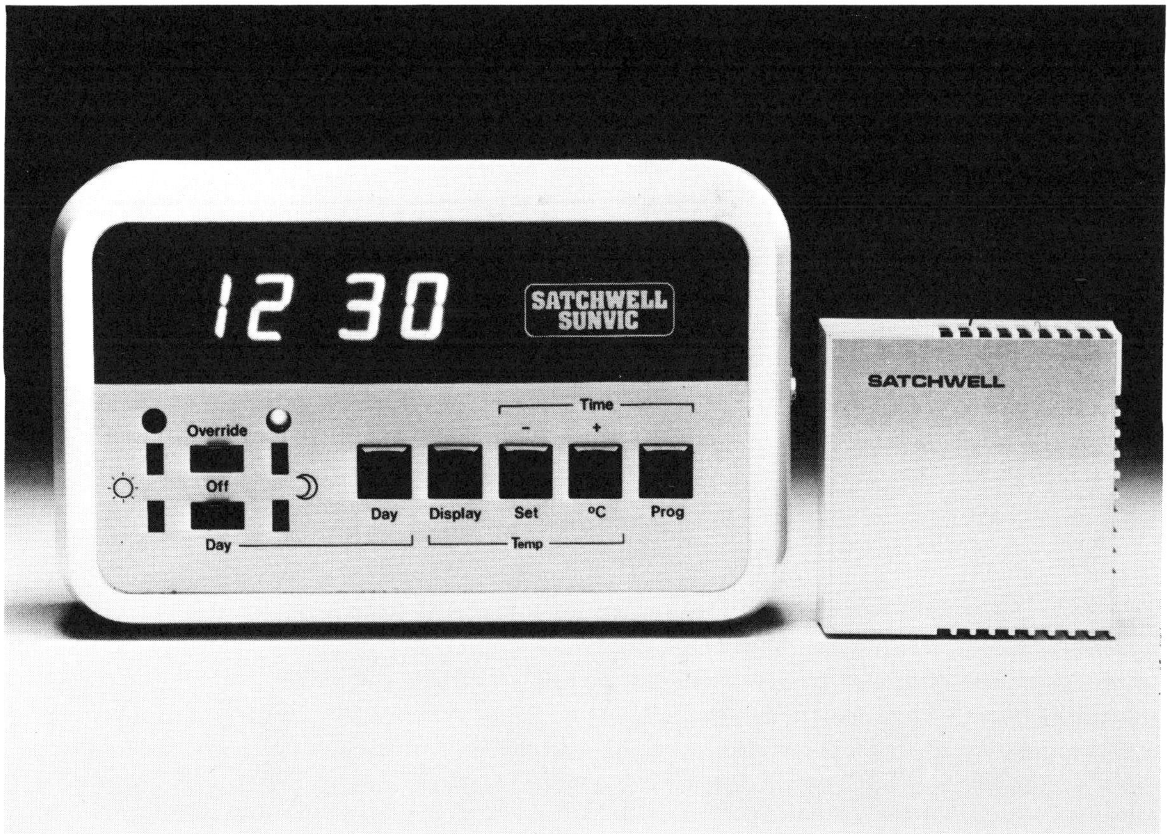

Source: Satchwell Sunvic Ltd

Figure 3.25: A more sophisticated heating control unit (the Economatic from Mayfair Electronics) with self-learning optimum start/stop facility.

Source: Mayfair Electronics Ltd

many of which are air-conditioned. The technique also forms an important monitoring and control function for many comprehensive energy management systems.

Air-Conditioning Control

Despite our cool climate, there is a more widespread use of air-conditioning in the UK than is generally realised. This is particularly so in modern deep plan office or retail buildings which have a tendency to overheat (even in winter) from the effects of lighting and occupants' body heat. Air-conditioning is a particularly complex type of buildings service because it involves two-phase heat transfer (i.e. evaporation and condensing) with the associated difficulties of measurement and control. Further problems commonly arise from frequent oversizing of air-conditioning plant (it is normally designed to cope with the hottest conceivable day with some remaining capacity to spare). As a result, compressors run most of their lives at well below full load and consequently well below peak efficiency. In addition, the design of the air circulation and heat transfer systems often results in simultaneous cooling of one part of a building (e.g. the sunny south side) and heating of another (e.g. the shady north side), with no direct transfer of heat between the two. Sometimes, one can even find heating and cooling going on in a single room! The net result is low overall efficiency and high running costs.

The poor design of many air-conditioning systems stems in part from their being constructed before the 1973 oil crisis at a time when minimum initial installation cost was paramount over running costs. The complexity of this subject makes comprehensive treatment of it inappropriate here. Furthermore, the scope for improving the efficiency of existing poorly designed systems by retrofitting improved controls is rather limited. Instead, I will confine my remarks to illustrating two particular control opportunities, one of which has been independently assessed under the ECDPS.

The first technique concerns optimisation of the temperature of the chilled water supplied from the refrigeration plant. Normally, this is kept constant at around 5°C regardless of cooling demand which is satisfied by starting and stopping of the compressors according to the temperature and humidity control set points. A microprocessor-based control unit known as the Chillitrol II and imported from the USA by Service First International of London is able to monitor cooling demand and allow the chilled water temperature to float up to an optimum level, which may be as high as 10°C. In theory, this should increase the coefficient of performance of the compressors and reduce electricity consumption by minimising the load on the electric motors. Additional functions such as compressor scheduling (i.e. minimising the number of compressors in operation) are also provided. Electricity savings are expected to be highest in winter when cooling demand is at its lowest.

A unit of this type (see Figure 3.26) has been installed by Norwich Union Assurance Co in their 7,000 sq metre All Saints Green office wing in Norwich with ECDPS support. The independent monitoring report shows that some electricity savings were made (around 17 per cent of previous consumption) but overall the financial return on the investment was poor with a payback of 36 years on the £14,500 capital/installation costs. However, the report (which is available from ETSU) does identify the types of circumstances where a more attractive return could be obtained, namely:

–in a larger building

–where air-conditioning is operated in winter as well as summer (the Norwich Union operation is summer only, and

–where the refrigeration plant is closely matched to the building cooling demand (it was rather oversized at Norwich).

The second important aspect of retrofit air-conditioning control is so-called enthalpy economisation. This technique enables the optimum use to be made of outdoor air for internal cooling (within the constraints of humidity set-point) and minimises the use of artificial cooling.

The complexity of the required control algorithms are such that this technique is normally only offered as one of the control functions of a comprehensive energy management system rather than as a dedicated control unit. However, as with other aspects of air-conditioning control, the accuracy, sensitivity and reliability of presently available humidity sensors remains a limiting factor on the performance and cost of this technique.

Figure 3.26: Chillitrol II air conditioning control unit. This optimises chilled water temperature according to building cooling demand.

Source: Service First International

The Energy Savings Potential from Dedicated Controls

The gross energy savings likely to be achieved by an assembly of advanced dedicated controllers of the types described in this chapter will depend on many factors such as the previous standards of operation and maintenance of plant and conventional controls. However, it is reasonable to assume that savings in the region of 10-20 per cent of the previous total energy costs can be saved on a typical building or site by applying them instead of conventional practices. Whether these savings can be maintained and improved upon over the long term is another matter and it is in this regard that the comprehensive energy management system begins to come into its own.

Chapter 4

COMPREHENSIVE ENERGY MANAGEMENT SYSTEMS

The Drawbacks with Dedicated Control Devices

We have seen in the previous chapter that there is substantial potential for reducing energy costs in buildings through the judicious application of dedicated control devices. A sound energy efficiency improvement campaign might include investment in some of the latest advanced microprocessor-based products. Further benefits can arise because the prospective user is able to shop around rather than being tied to a single manufacturer. He can also proceed at his own pace, installing one type of system at a time and learning as he goes along.

Nevertheless, there can be distinct disadvantages in employing dedicated control units rather than a comprehensive energy management system, and the prospective user needs to be aware of these. For example a clash of control strategies might arise with an assembly of dedicated control devices, for instance between space heating and air-conditioning control. It is easier to avoid this sort of problem with comprehensive systems which provide a central supervisory facility.

In other cases it may be hard to keep track of what is happening to a collection of dedicated controllers scattered around a large building or site. This difficulty becomes compounded when multiple buildings located around the countryside (e.g. schools) are involved. Units may malfunction or settings may be tampered with and undetected energy wastage may occur for months before the defects are discovered and rectified. A typical instance would be a school headmaster coming in to mark papers on a Saturday morning and, in overriding the normal programs, possibly causing heating to be applied 24 hours per day, seven days per week until the caretaker notices what is going on. More serious failures may have to wait until one of the infrequent visits by a council engineer before the problem is rectified. Thus, although dedicated controllers are capable of achieving substantial savings in the first instance, it may not be easy to maintain these savings over a long period. The remote monitoring and control capabilities of the comprehensive energy management system should, therefore, contribute substantially to the successful maintenance of energy efficiency.

A further consideration is that a disaggregated assembly of dedicated control devices may end up costing more to purchase and to operate than a comprehensive energy management system carrying out a similar range of control functions. This may be particularly true of manpower costs if large numbers of individual dedicated control units are involved.

Finally microprocessor-based energy management technology is capable of providing substantial amounts of useful energy management information, e.g. on energy consumptions, plant parameters, comfort conditions etc. With an assembly of dedicated control devices, this information is unlikely to be easily accessible and made full use of in the ways that are possible using the centralised input/output facilities provided by a comprehensive system.

Additional Features Provided by Comprehensive Energy Management Systems

A comprehensive energy management system can provide any or all of the monitoring and control functions described in the previous chapter, although it is not yet standard practice to include oxygen trim control of boilers. In addition, a comprehensive energy management system can provide the valuable centralised supervisory facilities mentioned in Chapter 2, namely:

–information input/output from a central station using computer-style peripherals such as visual display units (VDUs), colour graphics, plotters, keyboards and line printers. The contributions that these devices can make to improved energy efficiency lie in their roles as aids to more effective day-to-day management of energy, plant and services. Typical information received would be reports on space temperatures, plant status, energy consumptions and alarm conditions. This information would be difficult or costly to obtain in any other way

–off-line analysis of energy consumptions and trends, using some of the spare computing capacity that may be available with the central station. This is of particular value to the energy manager in his targetting, monitoring, forward planning and in the assessment of other possible energy conservation measures

–remote monitoring and control of plant, services and comfort conditions. This not only contributes to improved standards of operation and maintenance but also to more effective use of manpower by minimising the need for local site inspections and adjustments. For example, energy managers are placed in a far better position to quickly overcome the energy efficiency problems arising from local control units malfunctioning or from their settings being inappropriately adjusted.

The additional potential for energy savings over and above that obtainable from an assembly of dedicated control devices remains to be confirmed, but it is reasonable to assume that an extra five to ten per cent might be achieved. Firmer figures should arise in due course from the independent monitoring exercises being carried out on a large number of demonstrations of comprehensive energy management technology that are being supported under the ECDPS. Examples of these will be given later in this chapter.

Comprehensive energy management systems fall into two main categories, centralised systems and distributed intelligence systems. They will now be described in some detail.

Centralised Comprehensive Energy Management Systems

A centralised energy management system (Figure 4.1) is one that has all the system intelligence concentrated in a single Central Processing Unit (CPU). Information is gathered by an array of sensors, such as for space and hot water temperature, humidity, hot water or steam flow rate, electricity, oil and gas meters etc. This information is fed via a network of cables to the CPU. There it is stored, manipulated and analysed according to the operating software in order that appropriate control actions can be determined. Control decisions are relayed to appropriate actuators (electrical switches, valves, air dampers, etc) through another system of cables. Energy management information can be displayed on the usual output peripherals, and control parameters (temperature set points, heating program times, etc) can be adjusted by means of a keyboard.

The centralised type of energy management system was the first to appear on the marketplace in the early 1970s. At that time, these systems were looked upon more as building automation systems designed to enable the management to cope with the complexities of a very large building or site. Fire and security alarm networks were often also incorporated, and the expected financial return was usually more in reduced manpower requirements than energy savings. The central station would often bear a strong resemblance to a small computer (Figure 4.2). Overall capital/installation costs would be substantial (£100,000 plus), restricting the use of these systems to buildings with substantial energy bills, typically over £400,000 per annum if a payback time of under two years was to be achieved on energy savings alone.

Firms from outside the UK have remained in the forefront of the development of large sophisticated centralised comprehensive energy management systems for large buildings or sites. The USA (with firms such as Honeywell Controls and Johnson Controls) and Switzerland (with Landis & Gyr and Sauter) continue to dominate this part of the market. This reflects the pre-eminence of the USA in the general field of computer technology and the sophistication of the services of their many air-conditioned buildings and the progression of Swiss firms from clocks and watches to time-switches and building services control.

These large centralised energy management systems have now become well established in the UK and are commonplace on large buildings or sites. Prominent examples can be found on large industrial sites (e.g. motor works such as BL at Cowley and Vauxhall at Luton), public buildings (the Heathrow terminals), and major prestige office buildings (the Westminster Bank HQ tower in the City of London). In particular, substantial uptake has occurred in large (500 plus bed) hospitals, where there are probably more systems than in any other buildings sector; this reflects the complexity of the plant and services in hospitals, and the need to maintain stringent standards of comfort conditions whilst at the same time constraining operating costs to within tight limits.

Because these large centralised comprehensive energy management systems are now well established, their suppliers have gained substantial experience and there are now likely to be few problems with installation and commissioning. Furthermore, their after-sales backup in terms of maintenance and the capability for future expansion of the system is usually excellent. Substantial libraries of operating software are available to satisfy the likely requirements of most prospective users; however, this software is not always particularly 'user-friendly', and operating staff often need to be reasonably 'computer-literate' to make full and effective use of it.

Large Comprehensive Centralised Energy Management System in a Pharmaceutical Works

One of the first comprehensive energy management systems to be supported as a demonstration project concerned a Johnson Controls centralised system at the Hull pharmaceutical works of Reckitt & Colman. The site is not a particularly large one; the 1982 energy bill was £356,000, made up of £181,000 for steam, £87,000 for gas and £88,000 for electricity. Nevertheless, the building services are extensive and complex, reflecting the stringent environmental conditions demanded for pharmaceutical production. For example, there is a considerable amount of air-conditioning plant.

Reckitt & Colman carried out a large number of energy efficiency improvement measures between 1976 and 1980, before seriously considering the installation of a comprehensive energy management system. These previous measures included building fabric and pipework insulation, heat recovery from refrigeration and ventilation systems, improved conventional controls and staff education and training, and are estimated to have yielded total annual energy savings of £106,000 from a total investment of £107,000. These represent the most easily won savings, obtainable from fairly obvious and straightforward measures and the kind of campaign that any sensible similar organisation should have embarked upon. Further additional savings are not so easily or cheaply obtained. Furthermore, it is not always easy to maintain all of the savings of the past, particularly where these rely on some positive participation by ordinary members of the workforce.

Consequently, a comprehensive energy management system was near the top of Reckitt & Colman's shopping list for future energy efficiency improvement through capital projects. One

Figure 4.1: A centralised energy management system.

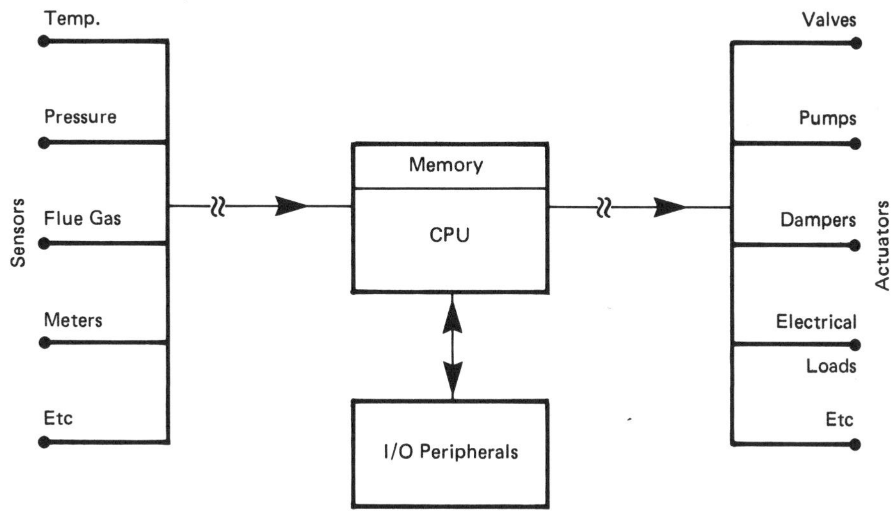

Figure 4.2: A typical centralised energy management system (Johnson Controls) central station.

Source: Johnson Control Systems Ltd

reason for this high priority was the complexity of their plant and the general difficulties of maintaining effective energy management. The equipment specification was sufficiently demanding for it to be most appropriately satisfied (when the order was placed in 1980) by one of the well-established centralised energy management systems of a type that had seen fairly extensive previous uptake in the building automation market. A smooth and reliable progress through installation and commissioning was important in order that production, quality or staff comfort should not be jeopardised.

The system finally selected (the JC 85/40) has evolved from the JC 80 series of building automation systems produced by Johnson Controls and aimed at the medium-to-large buildings market.

At Reckitt & Colman, the system comprises the following principal items of hardware:

Central Station (similar to the one shown in Figure 4.2) with:

–CPU (based on a Texas Instruments minicomputer)

–interface units to sensors, actuators and input/output peripherals

–memory (196k bytes of RAM plus additional PROM)

–colour VDU to provide both alpha-numeric and plant schematic displays

–keyboard and printer

–electronic clock and calendar.

Field Processing Units (FPU); 38 of these are installed around the site, and their purpose is to monitor and control sensors and actuators in their immediate vicinity according to the instructions received from the CPU. However, these FPUs cannot act independently from the CPU.

Sensors, including:

–external and internal air temperatures

–external and internal air humidities

–chilled and hot water temperatures

–heating and ventilation plant status

–electricity, gas and water consumption.

Actuators, including:

–electric-to-pneumatic transducers for operating valves and dampers, etc

–electrical switchgear, for lighting and ventilation fan control, etc.

The operating software of the system provides the following facilities:

–basic monitoring of energy consumptions and flows, comfort conditions and plant status

–centralised adjustment of control set points

–alarm indications (when monitored variables go outside pre-set limits)

–time/type of day programming of plant starts and stops, including holiday skips to take into account anticipated periods of factory shut-downs

–trend logs of principal variables, e.g. space temperatures, energy consumptions, boiler efficiencies

–enthalpy economisation for air-conditioning

—chilled water temperature optimisation

—electrical load management

—optimum start/stop of space heating

—pulsed reset control of lighting.

The total capital/installation cost of this system was £198,400, excluding in-house staff costs. Government support of 15 per cent of this cost was received under a Department of Industry Regional Development Grant and 25 per cent of the remainder was funded by the Department of Energy under the ECDPS in exchange for access to the project and the information arising from it. The cost breakdown is shown in Table 4.3.

It is interesting to note from these figures the high proportion (40 per cent) of total costs accounted for by work of a general engineering/installation nature, as opposed to hardware and software costs associated directly with the energy management system.

Annual financial savings (at 1982 prices) were expected to arise from the system and these are detailed in Table 4.4. They represent a crude payback time of just over four years on the investment in this project. This is substantially longer than the 12-18 month payback times typically demanded by manufacturing industry for energy conservation investments in general; this typical financial stringency stems from the currently low levels of real profitability of UK industry and the consequent shortages of investment capital. It therefore seems that relatively few other manufacturing concerns would have been prepared to go ahead with a similar project on the strength of this expected rate of return.

The project has been independently monitored since early 1981. Published reports are expected to become available from ETSU in mid 1984. Interim results suggest that the project has been a technical and financial success; relatively few problems were encountered with installation and commissioning, and the supplier, Johnson Controls, is regarded as a satisfactory contractor. Their performance is particularly creditable in view of the complex nature of the plant and services on this site.

Reckitt & Colman estimate that actual financial savings of £91,000 per annum are currently being achieved, although preliminary independent results suggest that this may be somewhat optimistic. Some of the savings have arisen directly from the energy management system; other additional savings have arisen from their using the system as a full and effective energy management tool, in particular to identify further areas requiring remedial actions. For example, it was discovered that the incorrect wiring and setting of existing frost protection thermostats was causing substantial unnecessary use of heating; such a problem would probably have remained undetected and unrectified had the energy management system not been installed. It also is now much easier to detect and remedy filter blockages than was previously possible. There are a substantial number of other indirect benefits arising from the use of this system. Figure 4.5 shows a typical schematic display that is generated on the VDU. This enables the factory engineers to rapidly assimilate and interpret substantial amounts of operational data and to make appropriate and effective energy management decisions.

Medium-Cost Centralised Comprehensive Energy Management System in a Medium-Sized Engineering Works

The type of large sophisticated centralised energy management system described so far in this chapter is clearly only appropriate to sites that have energy bills large enough to justify the capital cost of the necessary equipment. Furthermore, these sophisticated systems require substantial operator expertise if they are to realise their full potential as energy management tools. For the majority of sites with around 2,000 employees or less, not only will these large systems be far too expensive, but the necessary trained and motivated operating staff may not always be available. Something cheaper and simpler is obviously required.

Table 4.3: Cost profile of comprehensive energy management system installed at Reckitt & Colman.

Central station	£34,400	(17.3%)
Field equipment	£34,000	(17.1%)
Plant interfaces	£16,000	(8.1%)
Building modifications	£2,500	(1.3%)
Power supplies	£12,000	(6.0%)
Pipework	£15,000	(7.6%)
Installation (manpower)	£50,000	(25.2%)
Engineering	£11,500	(5.8%)
Inflation and contingency	£23,000	(11.6%)
Total	£198,400	

Table 4.4: Anticipated financial savings.

Start/stop time programming	£20,000
Holiday skip	£900
Improved maintenance	£2,400
Set point control	£2,700
Extended equipment life	£6,800
Lighting control	£3,000
Boiler scheduling	£6,200
Ventilation control	£3,700
Boiler/chiller monitoring	£1,800
Refrigeration control	£1,100
Total	£48,600 per annum

Fortunately, several cheaper, simpler systems have become available since 1980. Exploitation of the low-cost computing power of the microprocessor, coupled with intense competition and general interest in microelectronics, has resulted in a plethora of new energy management products aimed at small-to-medium sized non-domestic buildings. Happily, UK firms have been in the forefront of these developments and now dominate this sector of the market.

An example of the more recent generation of lower-cost centralised energy management system has been installed at the domestic central heating system manufacturing works of Potterton International in Warwick. The site comprises a number of typical production processes, including iron-founding, sheet metal forming and spray painting. A vigorous energy efficiency improvement campaign has been underway since 1979. This has included straightforward measures such as factory roof insulation, fitting of thermostatic radiator valves in offices, installation of a new high efficiency gas-fired heating system with heat recovery, extensive draughtproofing, together with general improvements to operations and maintenance, staff education and training, administration and policy. As a result, the energy bill on the 19,000 sq metre site has been reduced by about 28 per cent to around 68,000 GJ per annum, after changes in production levels have been taken into account.

As at Reckitt & Colman, the Law of Diminishing Returns began to apply to energy conservation investment, once the above measures had been completed. In order to achieve further savings, more ambitious investments with possibly lower rates of return became necessary. An energy management system was considered appropriate, not only to achieve further savings by improved control, but also to help maintain the existing savings through a monitoring capability. It was expected that this would also help Potterton's engineers to identify opportunities for future savings measures.

The system selected by Potterton was the 'Micropower 100', developed by the British firm Transmitton Ltd of Ashby de la Zouch. This carries out a more basic range of monitoring and control functions than the JC 85/40 system at Reckitt & Colman. The principal control functions are:

–optimum start/stop and set point control of the new heaters

–time/type of day control of the powder paint plant, wet arrestor ventilation/filtration system on the foundry, and the iron melting cupola fan.

The central station (Figure 4.6) on the system is also more basic than that on typical large centralised systems. A black-and-white VDU, a keyboard and a printer provide information input and output. It is microprocessor–rather than minicomputer–based, and data storage is provided by solid-state bubble memory. The station is linked by four-core cable to four outstations which provide interfaces for data gathering and control signal transmission. The system can display the following energy management information:

–historical space temperature over the previous 48 hours

–current status (on/off) of controlled plant

–cumulative hours run of time-controlled plant

–alarm messages where parameters exceed preset limits

–times of energy management system faults.

The total cost of the Micropower 100 system at Pottertons was £31,029, of which £23,549 was for equipment and £7,480 for installation work. Energy cost savings of around £13,000 per annum were anticipated, mainly on space heating costs but also on the operation of the items of process plant being controlled.

The project has been closely monitored for several years under the ECDPS. Accurate quantification of the actual savings has proved extraordinarily difficult because of the continuous changes to plant and production procedures that have occurred during the life of the project.

Figure 4.5: Typical schematic display generated by the Johnson Controls JC 85/40 energy management system at Reckitt and Colman's pharmaceutical works in Hull.

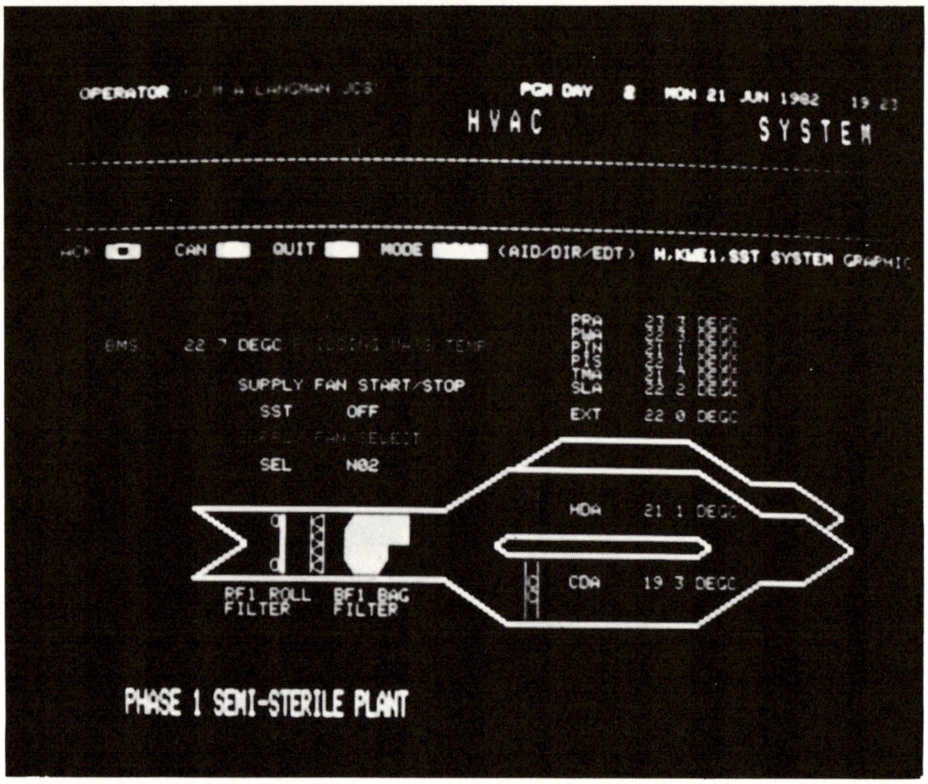

Source: Johnson Control Systems Ltd

Figure 4.6: The Transmitton Micropower 100 Central station used by Potterton International at their Warwick works.

Source: ETSU

Nevertheless, gas consumption has certainly not increased, despite a 15 per cent increase in production levels; more definitely, it is estimated that present consumption would have been 4-22 per cent higher if the energy management system had not been installed. The wide margin of error reflects the difficulty of separating the effects due directly from installation of an energy management system from all other factors such as climate, other energy efficiency improvements, and changes in working practices; these might have a considerably larger total effect. Attention focused on a small, relatively stable tool-room area has shown that the Micropower 100 is reducing heating consumption by around 13 per cent, mainly by tighter control of space temperature set points. Optimum start/stop is believed to contribute only three to four per cent.

The total value of these savings is estimated to be currently around £10,000 per annum giving a payback of just over three years. Again, this is somewhat longer than most manufacturing firms would like for energy conservation investments.

Despite the modest return on investment, this project is regarded as an overall success. An interim ECDPS report is expected to be available from ETSU in mid 1984. The main lessons are:

(i) Experience at Pottertons supports the view that energy efficiency should be raised to the highest possible level that is practical with well-established cost-effective measures before an energy management system is installed. The system will stand a better chance of providing real benefits if the plant and services it is monitoring and controlling are in good order.

(ii) The success of the project is believed to be due in part to the allocation of motivated and well-trained staff to operate the system. Pottertons paid particular attention to this point, and as a result have been able to exploit the system fully as a management tool.

(iii) However, the modest savings have largely been buried in the 'noise' of the site energy consumption data. Consequently, it would have been hard to have subsequently verified the expected savings if an extensive independent monitoring exercise had not been carried out at considerable government expense. Most future prospective replicators will not have the benefit of such an appraisal exercise, and will have to rely more on trust that the system is achieving its financial goals. Furthermore, it casts doubts on the credibility of many of the enthusiastic claims regularly made in the technical press regarding similar systems of this type!

Low Cost Comprehensive Energy Management Systems on Small Industrial Sites

As one moves down the building size range, further restrictions are placed on the justifiable scale and cost of an energy management system. Apart from the need to maintain a reasonable financial return, there may not be a specific member of staff available on a small site to take routine responsibility for the system. Many of the computer-style peripherals provided in the projects described above will not be used to the full. Consequently, a 'black box' approach is likely to be more appropriate, with data input/output facilities restricted to those normally found on a desk calculator, i.e. a keyboard plus a digital display.

Two projects demonstrating this simpler low-cost, more basic approach have been supported under the ECDPS. One involves the development and demonstration of the 'Enerite' energy management system by NEI Electronics at their Gateshead factory. Their premises (400 employees) are typical of modern small-to-medium sized sites, and are located on the Team Valley Industrial Estate. The system carries out a similar range of tasks to the Transmitton system at Pottertons but also includes lighting and maximum electrical demand control. The cost is considerably lower, at around £8-12,000 installed, and this is reflected in the much more basic facilities provided by the central station (Figure 4.7). However, a cheap printer can be supplied as an optional extra if required.

Nevertheless, despite its low cost and basic features, the system is quite powerful. It is also fairly sophisticated in its design, being based on the distribution of intelligence amongst individual remote control units (RCUs) carrying out discreet management tasks. This minimises the cost of installation and maximises the flexibility of the system with regard to subsequent modification and

Figure 4.7: The central station facilities on the low-cost NEI Enerite energy management system installed in the maker's factory in Gateshead.

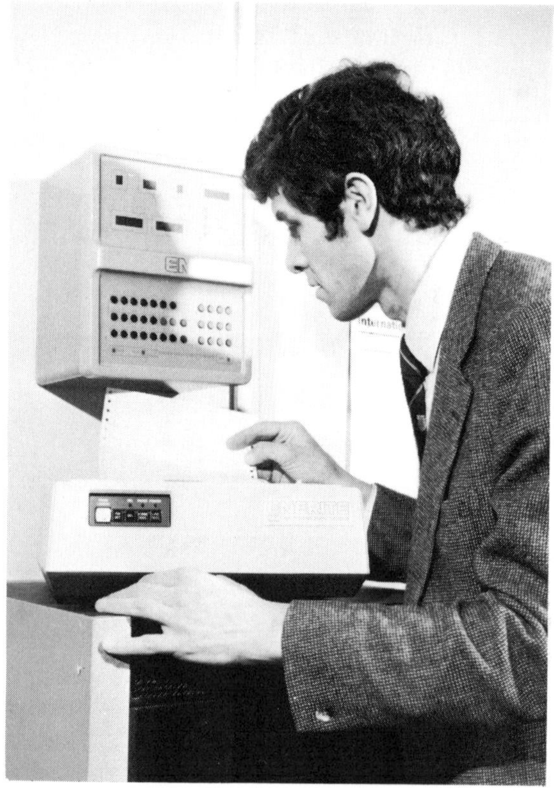

Source: ETSU

Figure 4.8: The central station facilities on the low-cost Energy Technology EM 2000 energy management system installed in the Gateshead factory of Crabtree-Vickers.

Source: ETSU

expansion. The RCUs are not capable of operating in a completely stand-alone mode; for example, they rely on the central programmer for the time and calendar information. Nevertheless, some of the control functions can continue even if the link with the central programmer (through the daisy chain data highway) is broken.

The development and implementation of this system has been somewhat prolonged; software development in particular has taken far longer than anticipated. This highlights the need for users to assess carefully the actual state of development of any equipment they propose to purchase; for example, it would not be very satisfactory to have a lot of expensive microprocessor hardware installed but be unable to use it until some missing software is delivered. At the time of writing the NEI Enerite was close to completion, and it will be monitored under the ECDPS over several heating seasons.

Another, somewhat similar system is also being installed in the nearby factory of Crabtree Vickers, also with ECDPS support. Their system is supplied by Energy Technology Ltd, a small Newcastle firm. It is known as the EM 2000 (Figure 4.8), and also employs a distributed intelligence approach to the system design. The total capital/installation cost to Crabtree-Vickers was just under £15,000. The equipment development and installation has proceeded fairly smoothly but quantification of actual energy savings awaits the results of an independent monitoring exercise currently being carried out. Interim reports should be published in mid 1984.

Remote Distributed Intelligence Comprehensive Energy Management Systems

All the comprehensive energy management systems so far described rely on a CPU in the central supervisory station to provide an effective alternative to conventional controls. This may cause problems if large numbers of buildings need to be connected to the system. For one thing, there is the need to consider what happens if the central station fails; it certainly would not be satisfactory for the occupants to go without heat and light while the problem was being rectified. A possible solution is to arrange for the system to fail-safe to the previous controls, providing of course that these are sufficiently sound to justify retaining them for this purpose. Secondly, as the numbers of buildings and plant rooms increase, so does the cost and complexity of the control and signal cabling. Finally, single CPU systems are totally inappropriate to groups of small-to-medium sized buildings scattered over a wide geographical area, particularly where individual buildings are too small to justify their own complete energy management system.

So whilst the centralised comprehensive energy management system concept might be appropriate for fairly compact sites such as factories, it is not necessarily the ideal solution for a disaggregated collection of buildings on a single site or where the buildings are dispersed over a large area, because of the costs of installation and the difficulties of ensuring continuity of control. Typical examples needing a different approach are universities, military bases and government laboratories on single sites, and schools, retail chain stores, hospitals, etc on dispersed sites.

Recent years have seen the introduction of the fully distributed intelligence energy management system. UK firms have been in the forefront of this development, and the dispersed buildings market has been uppermost in their minds. Figure 4.9 illustrates the concepts involved; instead of having all the system intelligence concentrated in a single CPU, it is distributed both geographically and in terms of the system control hierarchy. A 'local intelligent outstation' can be given the task of handling all the routine monitoring and control functions for an individual building or plant room on a day-to-day basis. This outstation contains all the computing capacity, memory and software to do this, but is unlikely to possess more than very basic data input/output facilities (e.g. simple keyboard and digital display). In this sense, it would almost be a complete energy management system in its own right, and may resemble one of the low-cost systems described in the previous section.

However, instead of interacting with a human operator, it is supervised by a central station similar to the type employed upon the large centralised systems. The link with the central station can be over a dedicated cable or an existing internal telephone network on a single site. However, the link can also be over the ordinary British Telecom telephone network, using automatic dialling

Figure 4.9: The distributed intelligence energy management system concept.

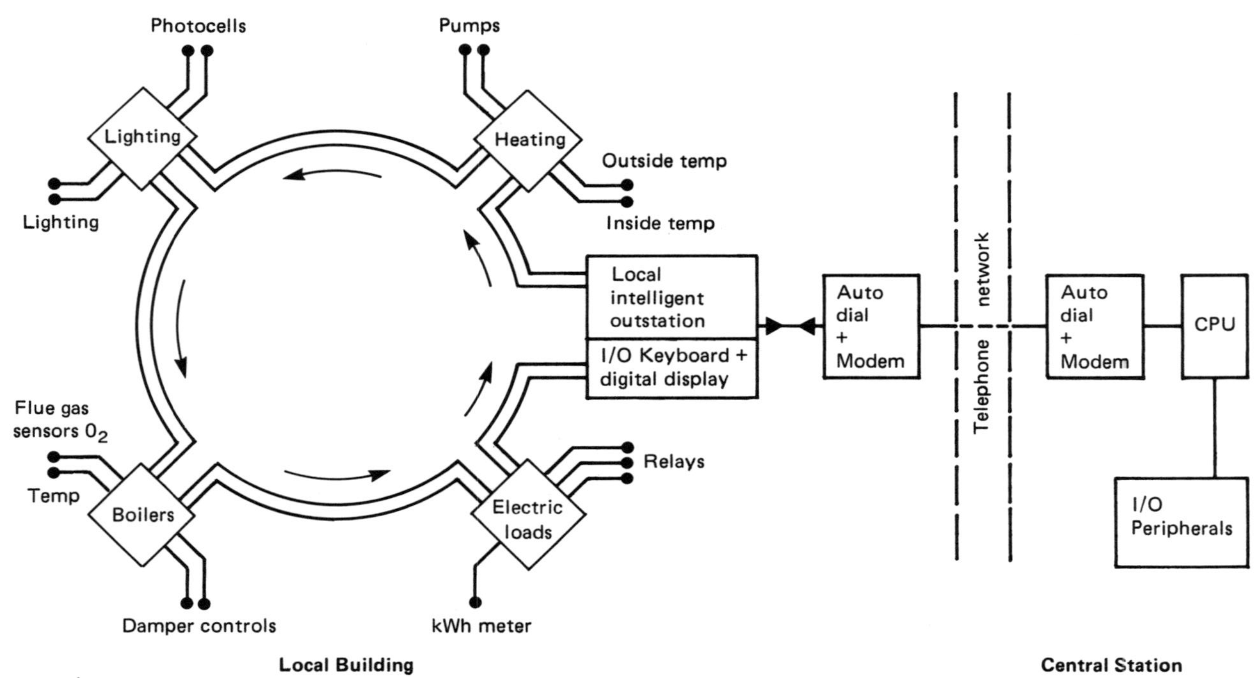

units and modems (signal modulation/de-modulation) to effect an electric/acoustic signal conversion. This allows the system to provide comprehensive energy management remotely to buildings many miles from the energy manager's office.

Furthermore, large numbers of outstations can be similarly connected, enabling the high cost of the central station to be shared among many buildings. Combining this advantage with the exploitation of low-cost microprocessor hardware for the local outstation means that individual buildings can be connected into a remote distributed intelligence energy management system for as little as £2-3,000 per building. This makes the technology potentially cost-effective for buildings with annual energy bills as little as £8,000, corresponding to a typical small supermarket or average sized primary school.

This distribution of intelligence can extend to the individual monitoring and control functions being carried out, with separate dedicated units covering lighting, heating and electrical load control (for example) under the general supervision of the outstation. The control hierarchy then reflects the typical management structure of an organisation, as shown in Figure 4.10. The control units represent the 'shop-floor' workers carrying out routine tasks (e.g. teachers in a school, nurses in a hospital). The outstation represents the local manager (headmaster, store manager, etc) who provides the workers with their day-to-day instructions. However, the outstation is programmed to transmit regular reports, for example on comfort conditions, fuel consumptions and any alarm messages if plant fails or parameters go outside pre-set limits, to the central station. The central station operator can then analyse and interpret the data received and send appropriate revised programming instructions to the outstation. In this regard, the central station corresponds to an Area Manager. Indeed, there may be more than one such supervisory station in a distributed intelligence energy management network, all linked to an overall supervisory station in a head office corresponding to the Chief Executive.

Distributed intelligence comprehensive energy management systems can compete effectively with centralised systems on many types of single site, especially if the number of buildings or plant rooms is large. Systems of this type are used increasingly on university campuses, large industrial sites and hospitals, etc. However, it is in linking groups of buildings dispersed around the country into a remote energy management network that the technology really comes into its own. It allows advanced monitoring and control techniques to be applied to buildings that are far too small to justify the costs of the technology available previously.

Because the distributed systems are relatively recent developments, they often have more advanced high level software in the central station, providing a more 'user-friendly' interaction with the operator than some of the longer-established centralised systems. Furthermore, they have a high degree of operational integrity because no major part of the control network can be disabled by the failure of any one component. Finally, the essentially modular nature of these systems enables the user to implement them in a 'building block' fashion, starting possibly on a pilot scale (with only a few outstations); if successful, it can subsequently be expanded in numbers of buildings, control functions and inclusion of a central station, to form an integrated network.

The demonstration of these distributed intelligence techniques to the remote energy management of various categories of small-to-medium sized buildings under common ownership has been a key feature of the ECDPS. A substantial portfolio of projects has been built up and much experience gained. Some of these projects and the lessons learned from them will now be described.

Remote Distributed Intelligence Energy Management Network in a Group of Supermarkets

One of the early pioneers of this technology was J Sainsbury Plc; in 1980 they embarked on a pilot project involving five typical supermarkets with local intelligent outstations connected by telephone to a central supervisory station in their Uxbridge area offices. The company arrived at this decision following a substantial programme of energy efficiency improvement measures, including the replacement of individual refrigeration compressors in cold display cabinets with centralised air-cooled refrigeration plant, and recovering the heat for space heating. Lighting load had been reduced by a third in some stores by removing every third row of fluorescent tubes, while in new

Figure 4.10: A typical energy management system hierarchy with its man-management equivalent.

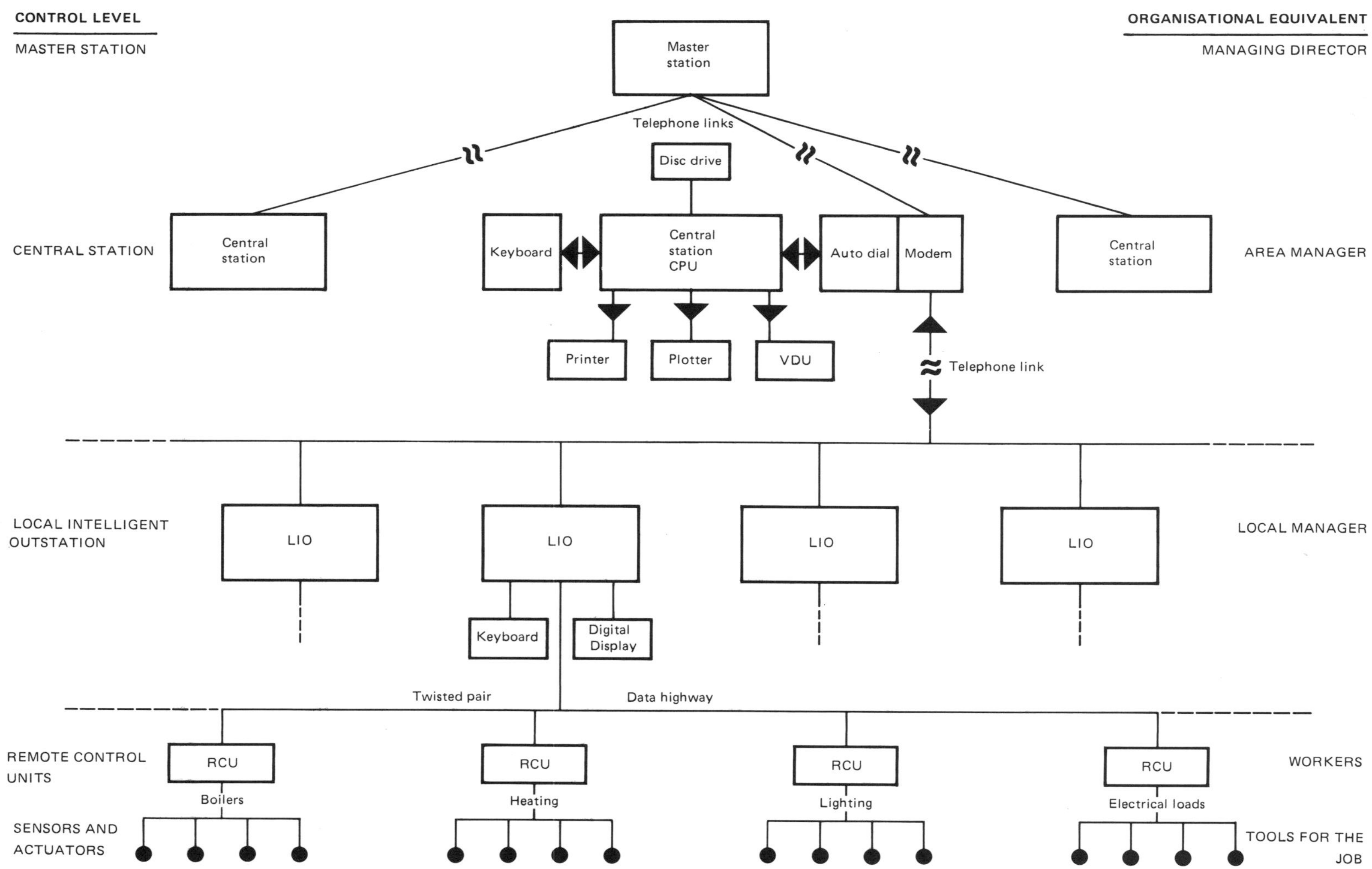

stores high-efficiency fittings have been used. These measures were taken against a background of major expansion in the number of stores, and the inclusion of established low energy designs in their new buildings. As a result of these measures, the company believes that its annual energy bill is about £1 million a year less than it would otherwise have been and hopes to achieve a further £1 million annual saving with the eventual widespread application of improved controls.

These savings may seem small in relation to the £2.3 billion annual turnover of the company, but they represent a significant fraction of the £100 million gross annual profit and are well worth striving for. The energy costs of a typical store are around £40-50,000 per annum and are dominated by electricity costs due to the large installed refrigeration and lighting loads. Energy management at the local store level has always been difficult because of a lack of suitably qualified staff and the preoccupation of store managers with pressing day-to-day commercial matters, such as maximising the turnover. Consequently, control settings are often tampered with and malfunctions in plant and services are not always promptly reported to the area office. It was appropriate, therefore, for some of this responsibility to be transferred directly to the area office by means of a remote monitoring and control system.

At the time that the project was embarked on, the choice of suitable equipment was much more restricted than it is now. In fact only one firm, ISS Compuguard, was able to meet their demanding specification. The total cost of the project has been around £100,000, with the central station accounting for about £20,000, the five outstations for around £40,000 in total, and local installation, sensors and meters, consultancy and management overheads for the remainder.

The system carries out a fairly standard range of energy management functions, including automatic time-based on/off switching, optimum heating start/stop, electrical load shedding and cycling etc. In addition, the central station is provided with information on air and water temperatures, plant status and electricity consumptions. Some limited information input/output capability is available at the outstations, but the main interaction is intended to be with the central station.

This project has had a difficult history and, to some extent, J Sainsbury has paid a price for pioneering the technology. Most of the problems have been of a technical nature, and stem from the ISS Compuguard equipment being far less well developed than the company or its advisors had realised at the time the equipment selection was made. ISS Compuguard (a joint USA/Dutch enterprise) was very vulnerable to hardware changes by its US suppliers; these required substantial and frequent software modifications, most of which needed to be carried out by subcontractors in the USA. The automatic telephone links were harder to effect than anticipated, especially through the British Telecom auto-dial/modem units which were eventually replaced by Dynamic Logic units, once the restrictions on the use of non-BT equipment were lifted in August 1982.

The inevitable result of these difficulties was substantial slippage in the timescales of installation, commissioning and utilisation. At one time, the project was 18 months behind schedule, and even now one of the important aspects of the specification, namely the ability to download outstation operating software from the central station, has not been implemented.

The problems with this project have caused considerable frustration, but the system has now been handed over by ISS Clorius (the re-organised successor to ISS Compuguard). The Sainsbury staff state that they are now well pleased and are making substantial use of the system as an energy management tool. The project is being independently monitored under the ECDPS, but no clear picture of the energy savings has yet emerged, partly because of the greatly extended installation and commissioning period and partly because of the continuous programme of other improvements and modifications to the stores and changes in business volume. These have masked any effects due to the energy management system. Certainly, total energy consumption has not gone down, but savings might emerge once a careful analysis has been made of the data to take account of all the other influencing factors. An interim report is expected to be available in mid 1984.

In the meantime, despite the difficulties encountered, J Sainsbury is convinced of the contribution that this technology can make towards minimising the operating costs of its business. Indeed, it is currently planning to extend its use to all new stores together with a limited number of existing stores. The experiences and expertise gained on this pilot project will stand it in good stead and provide an advantage over its competitors, who are only just beginning to embark on similar schemes.

Remote Distributed Intelligence Comprehensive Energy Management Systems in Groups of Schools

Schools are a prime area for the application of distributed intelligence energy management techniques because of the large numbers of sites under the control of individual educational authorities. In addition, the total energy costs are high and there are inherent problems in maintaining reasonable standards of operations and maintenance, good housekeeping and energy efficiency in this type of building. Headmasters have their minds understandably concentrated on matters of education and general management. Teachers are only too keen to order windows to be flung open to allow ingress of healthy fresh air without always realising the energy consequences of their actions. Caretakers often lack the knowledge and training needed to be able to maintain required comfort conditions at minimum energy cost.

Many local authorities have invested heavily in improvements to building fabric and plant together with better dedicated heating controls. As a result, energy consumption is a lot lower than it would have been only a few years ago. They have been helped in part by the fact that the rate of financial return for such investments does not have to be so high in the public sector as it does, for example, in private manufacturing industry. Nevertheless, there is still much more that could be achieved and it is not always easy to maintain the savings that have been won to date. As mentioned earlier in this chapter, advanced dedicated heating controllers quickly get tampered with by over-enthusiastic staff and the correct settings possibly lost. Furthermore, it is not always clear whether the regular maintenance procedures supposedly being carried out by contractors are being done properly or done at all.

Remote monitoring and control of heating plant in schools could, therefore, offer considerable scope for improving and maintaining energy efficiency. However, the actual realisable potential remains uncertain, bearing in mind practical constraints and the variable nature of school buildings, fabric and services.

The schools sector has therefore been a major area for demonstrations under the ECDPS. A total of five projects have been supported under that scheme, with a sixth receiving funds under a similar scheme run by the EEC. The principal features of these six projects are given in Table 4.11.

Table 4.11: Details of schools projects supported under the ECDPS.

Council	*Equipment Supplier*	*No of Schools*	*Total Capital/ Installation Cost*
Grampian Region	Atlantic Instruments	9	£260,000
Hereford & Worcester County	Transmitton	50	£260,000
Staffordshire County	Transmitton	3	£18,000
Essex County	Freeman Enercon	20	£250,000
Kingston on Thames Borough	JEL	10	£36,000
Somerset County	Transmitton	1	£63,000

These six projects differ substantially in scale, content and complexity. Nevertheless, they all comprise a central supervisory station in the Council Offices linked to outstations, sensors and actuators in individual schools and colleges. All the projects involve equipment by British suppliers.

The Grampian Regional Council project is probably the most ambitious and sophisticated of the six. In particular, it carries a high level of monitoring, not only of comfort conditions and plant status but also of external weather conditions and energy flows (Figures 4.12-4.15). The rationale behind the Council's thinking is that an energy management system cannot justify the name unless it provides a full set of energy information on which management decisions can be made. For example, a school may become cold for several reasons; the heating plant may have failed, the control settings may have been tampered with, windows or doors may be left open or external weather conditions may be exceptionally severe. The appropriate actions by Council engineers will obviously be different in each case. Furthermore, the engineers will not be able to assess the state of plant unless they have substantial information on its operating conditions. Finally, it is not always easy or possible to resist requests for increased heating from headmasters unless evidence can be provided that existing temperatures are adequate.

The system chosen by the Council was developed by Atlantic Instruments of Aberdeen to meet the Council's particular specification. Each school is provided with a substantial number of sensors for air and water temperatures, gas and electricity meters, oil tank levels, etc. In addition, several schools have automatic weather stations which provide information on wind speed and direction, solar intensity and rainfall. All this copious information is fed back by automatic telephone links to the central station (Figure 4.16), which needs to be provided with colour VDU and flatbed plotters to enable swift visual interpretation of the data. Figure 4.17 shows a typical schematic presentation of 24 hours feedback from one of the schools.

Even with the existing facilities, and only nine schools, it is easy to generate a 'data diarrhoea' situation, and further software may be required to provide adequate off-line analysis of the information provided. This situation will intensify as the numbers of schools connected to the system is increased from the present pilot scale to include more of the 300 or so premises under the Council's control.

Quantification of savings has been complicated by slippage in the timescale of installation and commissioning of the energy management system, and by other improvement programmes and changes being carried out on the schools involved in the project. A first interim monitoring report should be available from ETSU in the near future, but actual identifiable energy savings are not likely to be dramatic. This partly reflects the practical difficulties of measuring the effect of an energy management system on total energy consumption but also suggests that the scope for savings is actually rather limited, once the more obvious energy efficiency defects have been attended to. Nevertheless, this system will probably eventually justify itself as a valuable management tool, enabling the Council to optimise the use of its limited resources. However, clear financial justification is going to be more difficult to obtain.

A separate behavioural study carried out has revealed a distinct lack of enthusiasm among school staff for the new control arrangements. Much of this stems from the fact that comfort conditions were considered barely acceptable beforehand, and are now perceived to have worsened marginally. The trouble is that the thermal characteristics of the schools involved in the project are still rather poor, despite the remedial improvements carried out to date. The situation is exacerbated by inadequate heating systems, and results in comfort conditions being quickly affected by the severe weather conditions which can often occur in these northern regions. The staff feel that the amount of unnecessary consumption of energy that needs to be eliminated is very small.

Furthermore, changing patterns of occupancy and usage of the buildings mean that heating requirements can vary considerably in both space and time. It has proved more difficult than expected to cater for these under centralised control, and reversion to manual arrangements has been necessary on a number of occasions.

Finally, communications between school and Council staff is regarded as inadequate. Very few teachers received any information about the new system, what it was meant to achieve and how it was to be operated. This heightened feelings of resentment about some of the limited control over heating that was previously available being removed from the local staff. Also, introduction of the new system has greatly complicated the procedures for lodging and dealing with complaints about comfort conditions; under the old arrangements, a direct approach to the caretaker could be relied upon to sort most problems out. Staff now have to depend on the Council engineers being alerted

REMOTE FUEL CONSUMPTION AND ENVIRONMENTAL CONDITION MONITORING WITH THE ATLANTIC INSTRUMENTS ENERGY MANAGEMENT SYSTEM INSTALLED BY GRAMPIAN REGIONAL COUNCIL

Figure 4.12: Gas metering.

Source: Grampian Regional Council

Figure 4.13: Wind speed and direction.

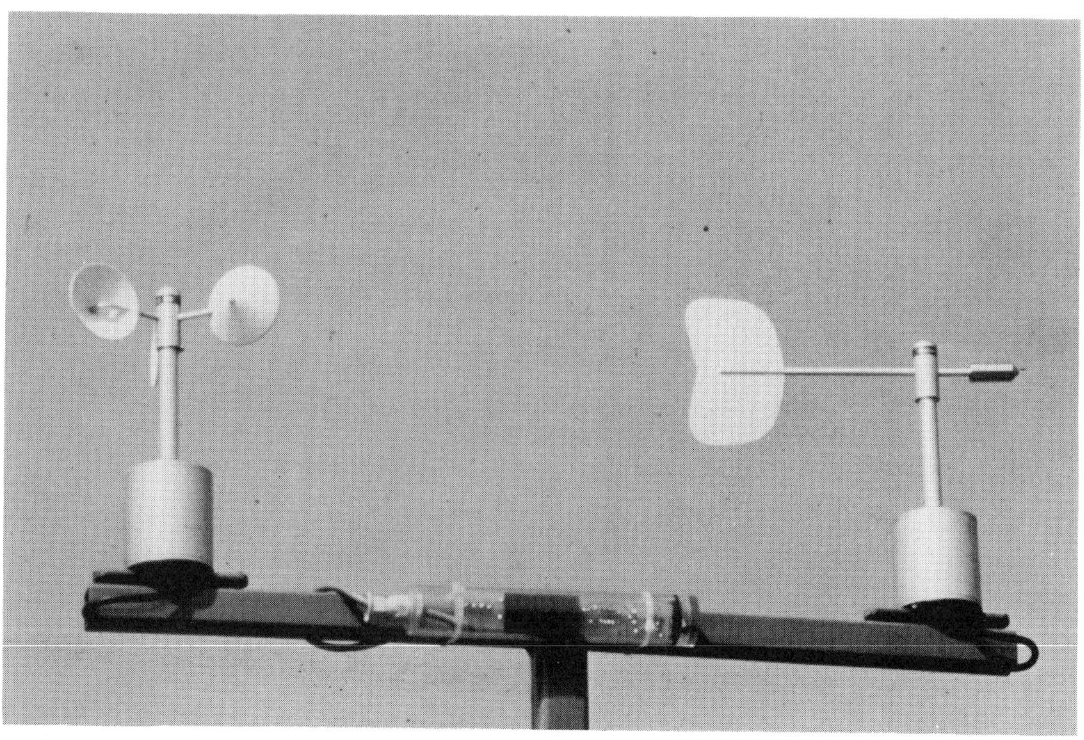

Source: Grampian Regional Council

Figure 4.14: Solar intensity.

Source: Grampian Regional Council

Figure 4.15: Rainfall.

Source: Grampian Regional Council

Figure 4.16: The Atlantic Instruments central station in the Council offices.

Source: ETSU

Figure 4.17: A typical schematic generated by the energy management system to show a typical 24 hours data collection in a Grampian Regional Council school.

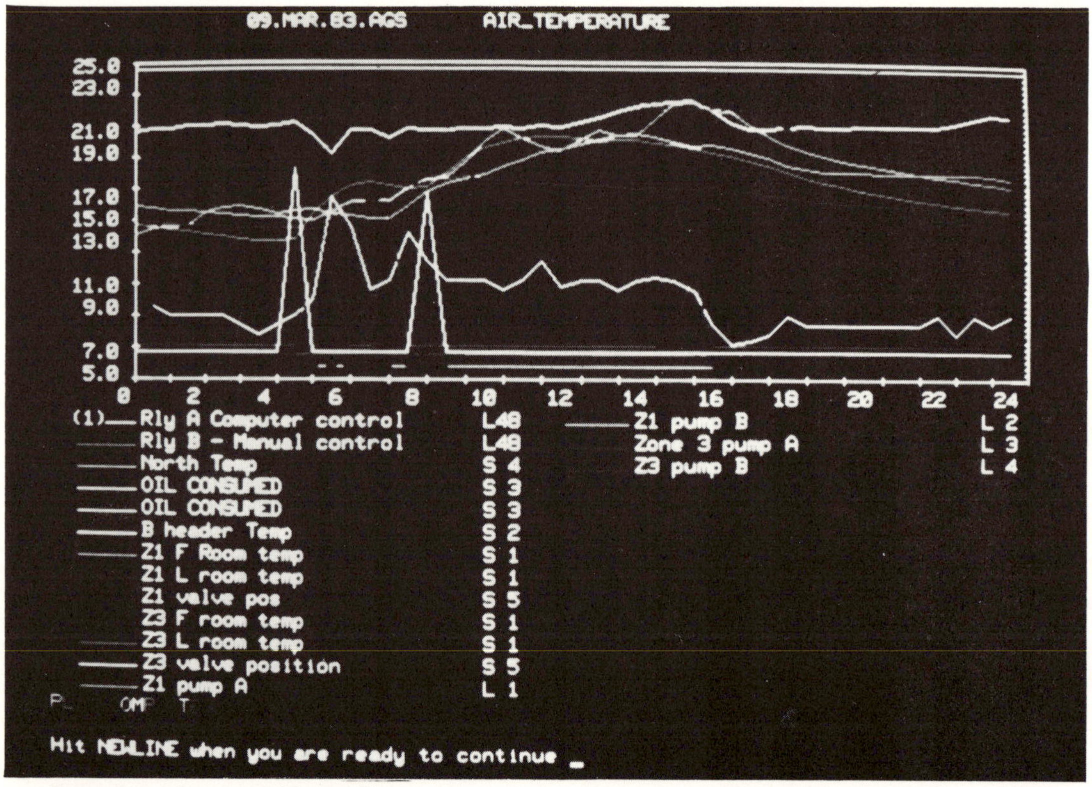

Source: Grampian Regional Council

to alarm messages at the central station, identifying the correct cause of the problem and initiating appropriate remedial actions. In practice, this appears to produce unsatisfactory results, partly because of the volume of information being received at the central station and the fact that the output is normally inspected only once a day.

The experience with this project illustrates the difficulties of simultaneously satisfying different requirements; local authority engineering staff are responsible for the buildings and the maintenance of adequate comfort conditions at minimum cost, but teaching staff have to live in the environment provided and put highest priority on the educational and physical welfare of the pupils. The two groups view energy usage from quite different standpoints, and considerable skill, tolerance and understanding is going to be required from the engineers if they are to successfully implement energy management technology.

The high costs of meeting the Grampian Regional Council specification mean that their approach is only appropriate for the larger secondary school or technical college. It also represents a higher investment per school than many other local authorities are likely to sanction in the first instance. The Hereford and Worcester County Council project is clearly less ambitious, because a comparable sum is expended on a network of 50 oil- and gas-fired schools rather than just nine. The local costs per school are as low as £3,000 in some cases, which puts the outstation equipment within the reach of the larger primary school. The equipment chosen is the Transmitton Micropower 2000 system, which is a fully distributed intelligence version of the Micropower 100 centralised system previously described. Its capability, particularly for monitoring, is significantly lower than the Atlantic Instruments system described above. Nevertheless, all the standard heating control routines are carried out and basic comfort condition and alarm monitoring information can be obtained at the central station.

The timescale for the implementation of this project has slipped considerably as a result of persistent defects in the system electronics and a prolonged software development programme. Nevertheless, 26 schools were on line in mid 1983. Although the level of information detail being received from each school at the central station is low compared to that in the Grampian Regional Council project, the number of schools involved is much larger. Consequently, the Council staff here are also finding it hard to cope with the amount of data being presented and a need for additional off-line data processing software has become apparent. This illustrates the wisdom of proceeding cautiously on a pilot scale in the first instance, so that subsequent investment decisions on larger scale implementations can be made with the benefit of substantial practical experience.

The performance of the system is currently being monitored under the ECDPS. Once again, the energy consumption picture is clouded by other factors, and it has proved hard to see any real energy savings against the background of everything else that is going on in the schools. Nevertheless, an interim monitoring report shows that average reductions in heating bills of around 13 per cent have been achieved so far (Figure 4.18). This corresponds to a simple payback period of just over four years, or an internal rate of return of 7.5 per cent, when operating costs are taken into account. For Hereford and Worcester County Council, like many other users of distributed intelligence systems, the main benefits from their system are likely to be in terms of an improved capability to manage their energy affairs effectively and efficiently.

A behavioural study of this project has revealed less intense dissatisfaction than with the Grampian Regional Council project. This may be due in part to the milder southern climate which does not highlight the building defects so sharply as in Scotland. On the whole, the effect of the energy management system on the teachers has been small, although some definite lowering of space temperatures in winter has been noticed by a majority of staff. It is also widely felt that local control over the heating system has been diminished and this appears to have led to an increase in the use of window opening as a means of controlling comfort conditions. Most teachers are very dissatisfied about the level of information and consultation provided by the Council before installing the energy management system.

A similar Micropower 2000 system has been supported as a smaller demonstration in a group of coal-fired schools of Staffordshire County Council. The particular energy management problem with coal-firing is the need to optimise the time the plant is fired when idling (the so-called kindling periods) to prevent the fire going out. This used to be adjusted manually by caretaker staff, and

Figure 4.18: The effect of installing an energy management system on the mean annual fuel consumption in five Hereford and Worcester schools.

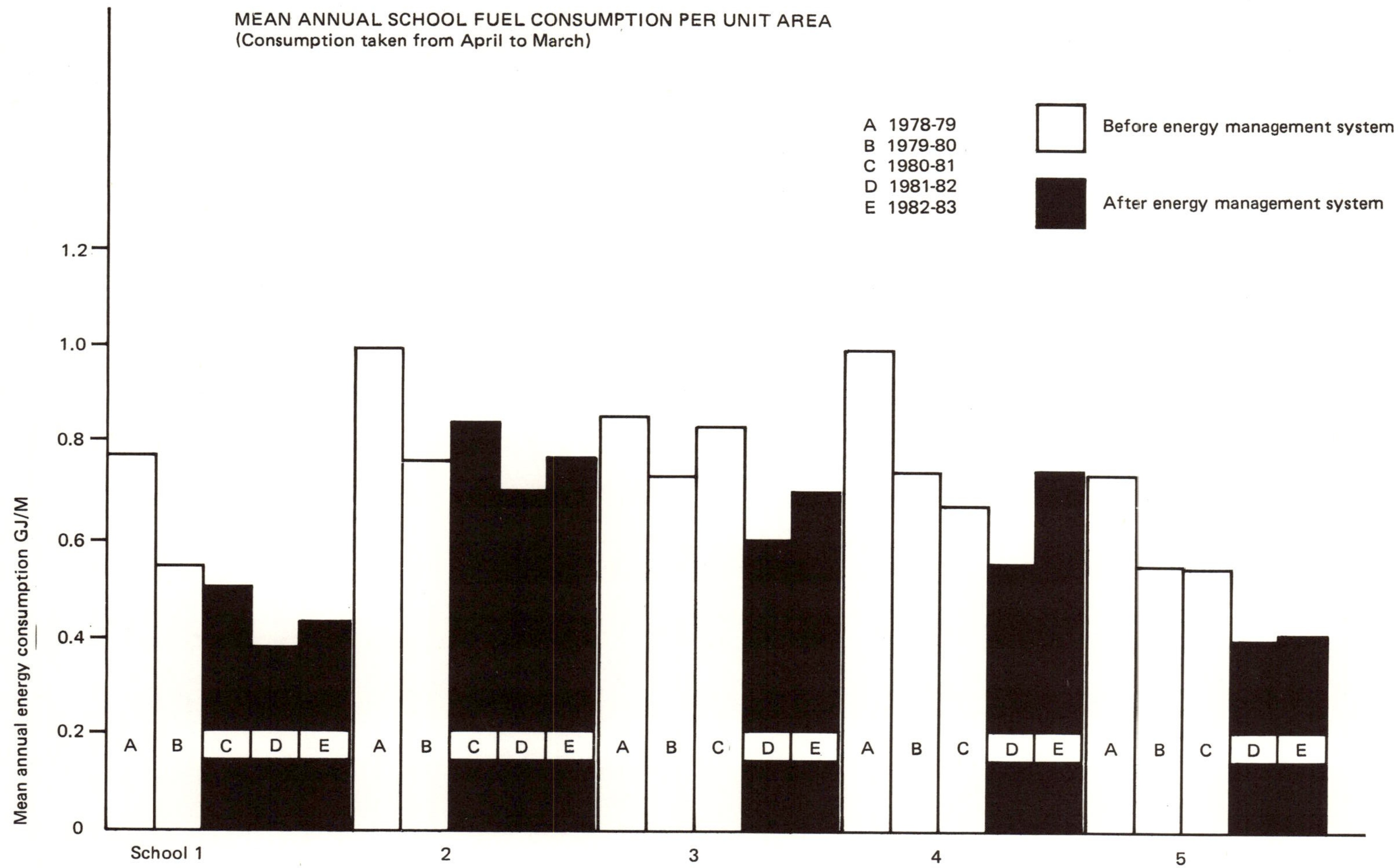

settings (and coal consumptions) varied considerably. The Council sought the capability to do this remotely, so that these times could be varied and optimised by their engineering staff. In addition, the system is intended to carry out the standard range of heating set-point and optimum start/stop control routines plus remote comfort condition and plant status monitoring.

In the event, progress with implementation of these requirements has been slow and remote kindling control has only been effected recently in the three schools included in the demonstration. The result has been that the independent monitoring organisation assigned to this project has not yet been able to report on the actual benefits. However, if the project is successful, it could prove very valuable to those local authorities in mining areas of the country who have a strong coal-burning policy, since there are special difficulties of monitoring and controlling the combustion of this fuel.

The Essex County Council project has received support under the demonstration scheme operated by the EEC. The Freeman Enercon 'Xyntax' distributed intelligence system is intermediate in sophistication, capability and cost between the Atlantic Instruments' system and the Transmitton Micropower 2000. Progress with installation and commissioning is believed to have been relatively smooth, although somewhat more prolonged than originally anticipated. At the time of writing, no information on actual energy savings has emerged from the independent assessors appointed by the EEC.

The Royal Borough of Kingston upon Thames have adopted a slightly different approach to the technology. Their distributed intelligence system is produced by JEL of Stockport, Cheshire and embodies a particularly high degree of modularity. Each of the individual energy management functions (e.g. for heating optimum start/stop, boiler sequencing) is carried out by a separate unit. This has enabled the Council to proceed using a 'building block' approach, installing a number of outstations to provide improved local control in the first instance. The central station and communications links to the outstations will be made at a later date to obtain the benefits of improved standards of operations and maintenance through the system's remote monitoring and control facilities. This project is presently in the stage of installation and commissioning of the outstations, and quantitative information from the independent monitoring organisation appointed under the ECDPS is yet to come through.

The sixth demonstration of comprehensive energy management technology in schools concerns a single educational complex in Taunton comprising a college of Art and Technology and a large secondary school. A Transmitton Micropower 100 system consisting of seven outstations connected to a central station has been installed on this site. It provides sequential control of the boilers, optimum start/stop and temperature set-point control of the heating zones and domestic hot water, two-stage frost protection and general monitoring facilities.

Installation and commissioning of the system is presently proceeding, albeit somewhat behind schedule. This project forms part of a major energy efficiency improvement campaign which has included replacement of two of the three central boilers to provide an improved match between heat supply and demand. Quantitative energy saving results are expected to become available late in 1984.

Distributed Intelligence Energy Management Systems in Groups of Hospitals

As stated earlier, the hospitals sector has probably seen the highest level of uptake of comprehensive energy management technology, with around 100 systems presently installed in the UK. However, the majority of these are large centralised systems in large single sites. Many of the small and medium-sized hospitals and clinics do not have sufficiently high energy bills or numbers of on-site engineers to justify one of these sophisticated systems all to themselves. Two projects have been supported under the ECDPS to demonstrate the sharing of the costs and facilities of distributed intelligence systems among a number of such sites.

One project concerns a group of hospitals of the Cornwall and Isles of Scilly Health Authority. The equipment involved is the Transmitton Micropower 2002, a more advanced and powerful version of the Micropower 2000 supplied to the Hereford & Worcester schools project described

Figure 4.19: The Transmitton Micropower 2002 central station at the Cornwall and Isles of Scilly Health Authority offices in Truro.

THE PLYMOUTH HEALTH AUTHORITY ENERGY MANAGEMENT SYSTEM PROJECT

Figure 4.20: The New Derriford Hospital in Plymouth. This is under the control of a Trend Controls energy management system which is also linked to three smaller outlying hospitals.

Source: Plymouth Health Authority

THE PLYMOUTH HEALTH AUTHORITY ENERGY MANAGEMENT SYSTEM PROJECT

Figure 4.21: The schematic arrangement of the distributed intelligence energy management system.

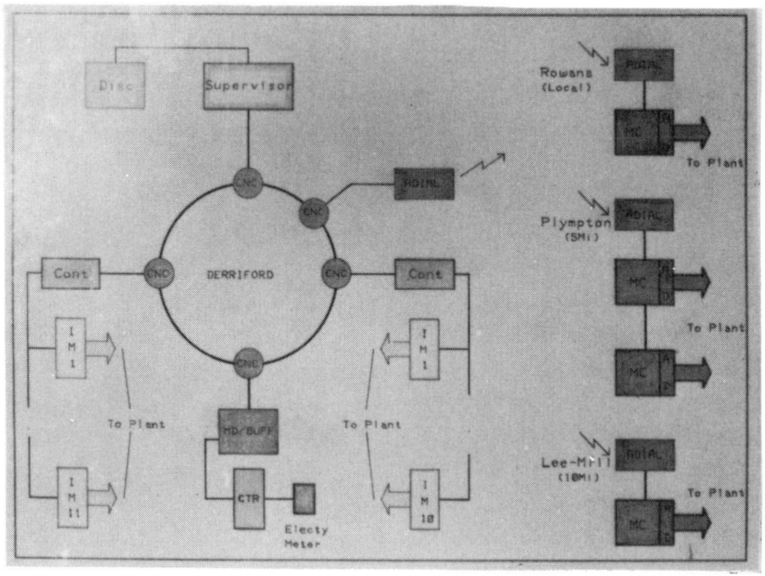

Source: Plymouth Health Authority

THE LADBROKE HOTELS ENERGY MANAGEMENT SYSTEM PROJECT

Figure 4.22: The JEL energy management system outstations in a hotel plantroom.

Source: ETSU

above. A central station (Figure 4.19) in the Truro headquarters of the Authority is linked to local intelligent outstations in six small outlying hospitals. It is also linked to a Micropower 100 central station that is monitoring and controlling the boiler house and various plant rooms on a large hospital site in Bodmin. The total capital/installation cost was £80,000.

The Health Authority hopes to achieve the normal energy cost savings expected from such a system through improvements in heating control and standards of operations and maintenance. In addition, they also hope to be able to make much better use of engineering and maintenance manpower, and to reduce the limitations previously imposed by the large geographical area covered by the Authority with roads often impassable through snow in winter and tourists in summer. The installation is proceeding against a background of other major energy efficiency improvement capital projects. These include conversion from oil to natural gas where this is available and eventual replacement of the central boiler house and steam distribution network at Bodmin with individual gas-fired heating systems in each of the many buildings on that site; this latter measure will greatly reduce the effects of heat transmission losses.

It is particularly difficult, therefore, for the independent monitoring organisation assigned to this demonstration project to separate out the effects of installing the energy management system from the combined effects of all the other measures being undertaken. Indeed, it has proved necessary to disable the new controls in one of the major buildings on the Bodmin site in order to simulate what the situation would have been if the new system had not been installed. Quantitative results are expected to emerge after the 1983-84 heating season. However, the Health Authority is claiming current fuel cost savings of £32,000 per annum giving a simple payback of under three years on the original investment. Furthermore, it is already actively expanding the system to include other hospitals, clinics and ambulance stations under its control.

The second hospitals project involving distributed intelligence energy management systems supported under the ECDPS is with the Plymouth Health Authority. It is centred on the new 625 bed Derriford District General Hospital (Figure 4.20) located on the outskirts of the city. This building has particular energy management problems stemming from the fact that it is presently well below its ultimate size but its heating and air conditioning plant is of sufficient capacity to meet all foreseeable expansion. In the meantime, much of the plant is oversized, and runs well below full load and peak efficiency.

The Plymouth Health Authority has installed a Trend Controls (formerly Holec Energy) 'Cadring' distributed intelligence system, comprising a number of local intelligent outstations on the Derriford Hospital site connected to a central supervisory station in the engineer's office. This central station is also connected by automatic telephone links to further outstations at three smaller outlying hospitals (Figure 4.21). The total capital/installation cost for the project was £53,000. The Health Authority hopes to achieve energy cost reductions of around £25,000 per annum. However, this is going to be hard to confirm conclusively because the Derriford site is a relatively new one and the previous energy consumption records are restricted to a few years during which the hospital was being brought into use and energy consumption was gradually building up.

A behavioural study of this project has identified many problems in common with the schools projects. In particular, the occupants' view of their comfort conditions is dominated by difficulties arising from defects and inadequacies in the building fabric and services. The effect of the new control system has been only marginal, but where a definite change has been noticed, the common view is that comfort conditions have been made worse rather than better. Again, lack of consultation by the Health Authority has not helped. Full details of the project should be available from ETSU as an interim report in mid 1984.

Distributed Intelligence Energy Management System in a Group of Hotels

A further example of distributed intelligence energy management techniques concerns a group of 15 hotels in the Ladbroke Hotels chain. The purpose of this project was to help overcome some of the energy management problems commonly afflicting such hotels, which usually have limited on-site engineering staff and hard-pressed managers who need to concentrate on other, more immediately pressing business matters. Customers must be kept satisfied at all costs, and to minimise the risks of complaints space temperature levels are often raised to well above the statutory limit of 19°C. The pattern of use of the various function rooms of the hotel may be very varied, and it is not always easy to match the supply of heating and ventilation to the utilisation of these rooms. Furthermore, plant and controls may malfunction for considerable periods of time before the faults are detected and remedied.

Ladbrokes invested in a JEL system (Figure 4.22) similar to that installed in the Kingston upon Thames schools project described above. They have also proceeded in a two-stage fashion, with installation and commissioning of the outstations proceeding ahead of that of the central station which is to be located in one of the larger London hotels in the chain. A fairly standard range of monitoring and control functions is to be applied to the heating and ventilation plant. The total capital installation cost for the whole project is estimated to be £165,000.

The project has so far progressed to the commissioning and handover of the outstations. Substantial delays were encountered during this stage with over six months slippage on what was expected to be a five month programme. It is now clear that the customer demanded and the equipment supplier promised a quite unrealistic timetable for implementation of the system, partly to meet the 1982-83 heating season. Some of the problems stemmed from teething difficulties and development delays with the JEL equipment. However, further difficulties were caused by institutional and organisational problems, particularly with regard to Ladbrokes' absorbing and utilising the technology. The hotel managers were not always as well informed about the details and purposes of the project as they might have been, and some were more inclined to override or disable the equipment than to learn how the new controls should be used. This was particularly so in the event of any customer complaint, whether it was due to equipment malfunction or more appropriate (i.e. reduced) heating control settings.

At the higher level, the necessary managerial and organisational adjustments have not yet been made to enable the remote monitoring and control facilities to be fully exploited. Responsibility for operating the central station as a remote supervisory facility has not yet been allocated and there is no proper authority to remotely adjust comfort conditions, possibly against local managers' wishes. The principal lessons arising from this project for other prospective users of similar technology are that an energy management system cannot meet its full energy cost reduction potential unless appropriate management authority and responsibility is allocated, and unless adequate information and training is provided to all those members of staff involved with its operation.

Distributed Intelligence Energy Management System as a Bureau Service to a Group of Tenanted Office Buildings

Our last example illustrates an interesting approach to the problem of energy efficiency improvement in tenanted buildings. It concerns a project with Energy Technique (Developments) Ltd, who operate what has become known in the trade as an energy management bureau (or a remote facilities management service). Briefly, this provides central supervisory services to clients through an energy management system on a fee basis. Sometimes, all the equipment is provided free of charge and the bureau takes a proportion (usually half) of agreed energy savings. This can be attractive because of the zero capital cost, but it can be difficult to agree what the savings actually are. Other arrangements involve the client carrying the cost of purchasing and installing outstations in his building, but paying the bureau for supervision from a remote central sation. Energy Technique offer the latter arrangement which has the advantage of a clearer understanding of the client's financial commitment. Other firms offering this type of service are Haden Maintenance, Welsmere Energy Management and CEMOC Ltd.

Use of a bureau service might be attractive to organisations who are reluctant or unable to enter into a full energy management system project of their own. This may be because they wish to try the technology out on a very limited scale in the first instance, or because they do not have the necessary in-house technical expertise to cope with a complete system. This is frequently the case

THE ENERGY TECHNIQUE (DEVELOPMENTS) BUREAU SERVICE ENERGY MANAGEMENT SYSTEM PROJECT

Figure 4.23: **The Trend Controls central station at the offices of Energy Technique (Developments) Ltd near Basingstoke.**

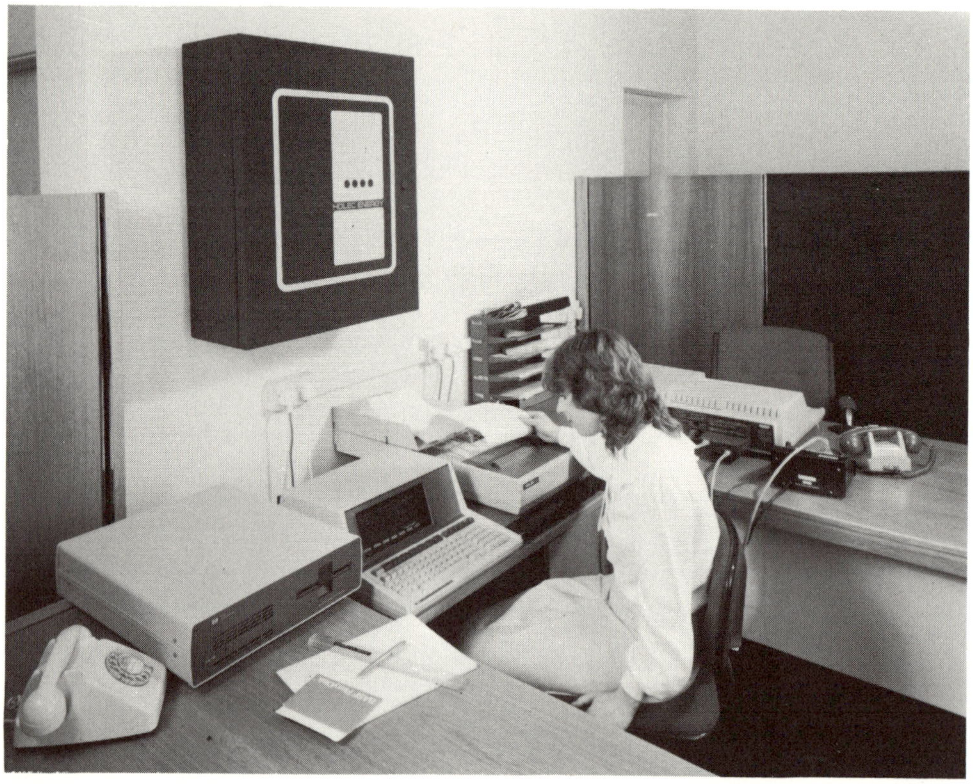

Source: ETSU

Figure 4.24: **A Trend Controls client's outstation linked to the Energy Technique bureau service energy management system.**

Source: ETSU

THE ENERGY TECHNIQUE (DEVELOPMENTS) BUREAU SERVICE ENERGY MANAGEMENT SYSTEM PROJECT

Figure 4.25: Energy savings achieved to date at York House. These average around 40 per cent of previous consumption.

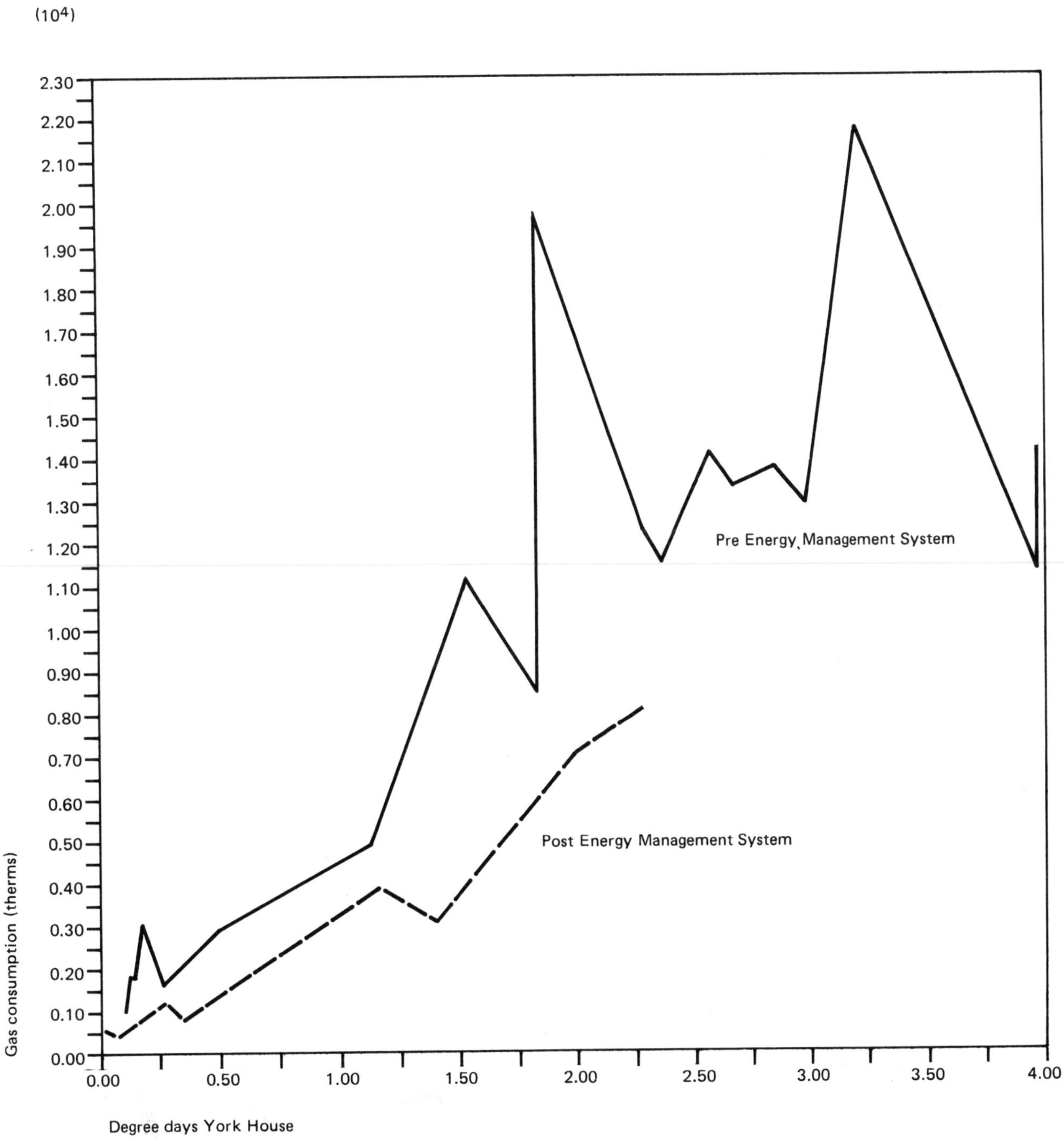

Source: ETSU

where tenanted buildings are concerned, because of the particular investment barrier that stems from the savings benefits accruing to the tenants (if they pay their own fuel bills) while the capital costs fall to the landlord. Using an energy management bureau minimises the capital and operating costs, whilst at the same time enabling the landlord to offer prospective tenants an energy efficient building.

Energy Technique are hosting a demonstration involving 10 tenanted office buildings linked into a Trend Controls distributed intelligence energy management system. The central station (Figure 4.23) is at the company's offices near Basingstoke. Figure 4.24 illustrates a typical client's outstation. Total capital expenditure has been about £154,000 with the central station accounting for £11,000 and the outstations for between £4,000 and £18,000 each, depending on complexity. The clients include a number of major property owning groups such as General Accident Assurance Co and the Church Commissioners. Close collaboration is maintained with the managing agents for the buildings.

Relatively few serious problems have been encountered during implementation of this project, although the programme has slipped somewhat from the original timescale. In some of the plant rooms, quite a lot of extra engineering work was found to be necessary to repair the effects of previous neglect before the new controls could be installed. Savings to date appear very promising, reflecting the substantial wastage that can occur in tenanted buildings. The system has already achieved reductions of between 15 and 40 per cent (see Figure 4.25) in energy costs in three of the buildings whilst at the same time maintaining adequate comfort conditions. Full details should be available from ETSU in mid 1984.

General Overview on Comprehensive Energy Management Systems

Several important lessons have emerged from demonstrations of comprehensive energy management technology. The first is that the effect of one of these systems on the energy consumption of a typical building is likely to be small compared to the combined effect of all other influencing factors, such as climate, normal variations in the patterns of building and energy use, and the implementation of established good operations and maintenance practices. This makes identification of savings achieved with energy management systems difficult. However, reductions of 10-20 per cent of heating costs do appear possible over and above those achievable with current good conventional practice.

Additional savings potential appears also to exist through the use of these systems as management tools, for example to identify scope for other additional improvement measures and to make the most effective use of engineering and maintenance resources. However, this potential has proved difficult to exploit, particularly on network projects involving multiple buildings connected into a remote distributed intelligence system. Indeed, the achievements of some present projects are seriously limited by the abilities of organisations to exploit the potentials of the systems they have purchased. There are a number of organisational, managerial, software and operator training problems yet to be overcome before they can realise the full potential value of this technology as an aid to more effective overall management. These problems are less severe in projects involving systems on single sites, and some encouraging progress is beginning to be made.

Reaction from building occupants to the installation of comprehensive energy management systems has so far been rather disappointing. Serious dissatisfaction has been revealed, particularly where distributed intelligence networks in the public sector have been involved. The principal cause has been the generally poor thermal characteristics of buildings in this sector, often combined with inadequate heating systems. This has given rise to frequent complaints about comfort conditions, which are perceived to have worsened rather than improved as a result of the new control technology. In many cases, lack of consultation and information has not helped occupants to understand the purposes and functions of the energy management systems that have been installed.

Few, if any, present comprehensive energy management projects can be regarded as complete successes. Most tell a mixed story of success and failure, but with much valuable experience being gained. The financial returns, where these have been quantified, appear modest, with most projects achieving simple payback times between three and five years. This is acceptable to public sector users, but is proving a disappointment to many in the private sector, particularly in manufacturing industry.

Chapter 5

THE ENERGY MANAGEMENT SYSTEM SUPPLY INDUSTRY AND MARKET

The General Character of the Equipment Supply Industry

As the real price of fuel has escalated in the last ten years, so a growing industry has sprung up to provide various means of improving energy efficiency and reducing costs. Some of the measures have been available for many years, but have only recently seen substantial market growth; an example would be low temperature insulation, such as glass fibre wool for roofs. Other measures, such as electronic energy management systems, only emerged once fuel cost inflation created the market and the necessary technology became available.

However, despite this dramatic growth in recent years, the energy conservation equipment supply industry in general remains small, disaggregated and weak. This is certainly true in comparison to the fuel supply industries; manufacturers of insulation or energy management systems, for example, pale into insignificance in relation to the size of the National Coal Board, the CEGB and Electricity Boards, British Gas or the major oil companies whose vested interests lie primarily in increasing their market share rather than decreasing the amount of fuel their customers use.

Furthermore, the energy conservation equipment supply industry invests relatively little on R & D in comparison to the fuel supply industries, who normally dedicate a much higher percentage of their turnover to this aspect of their businesses. The latter also allocate very substantial resources to marketing and promotion, an aspect often seriously neglected by the conservation industry. Finally, some conservation systems are not manufactured at all in the UK and the prospective user is obliged to import the required equipment; typical examples may be found in heat recovery, especially where heat pumps or special heat exchanger materials are involved.

The general picture of the UK energy conservation equipment supply industry is that it lacks strength and vigour; this is likely to be a major barrier preventing the available technologies from achieving their full long-term potential for reducing the national energy bill. Some members of the industry have banded themselves into organisations such as ACE (Association for Conservation of Energy) and ESTA (Energy Systems Trade Association) to obtain a greater influence on government and enhanced credibility in the marketplace.

However, there are some exceptions to the above picture and the UK electronic energy management system supply sector of the industry is one of the brightest. It is certainly very disaggregated (around 100 firms presently claim to be able to supply systems), and most of the participants are very small (less than 50 employees). Nevertheless, it is extremely vigorous and dynamic, partly because of its strong links with the booming microelectronics industry and partly because it is populated mainly by aggressive, young, ambitious technocrats, eager to work for themselves, to make a success of their businesses, and to exploit the technologies and market to the full.

The electronic energy management system supply industry has put in a very substantial amount of development work before placing suitable systems onto the market. Much of the hardware has been obtained from upstream major suppliers and is itself the product of substantial R & D (e.g. microprocessors and associated input/output peripherals). Relatively little hardware has had to be specially developed for energy management systems, and they often bear a strong visual resemblance to many other types of intelligent devices such as business computers, word processors and process controllers.

It has been in the application of this technology to the monitoring and control of buildings that the major developments have occurred. In particular, much work has gone into software development for operating the systems and providing the particular energy management applications. Indeed, the value of these vital components represents roughly half the value of a typical working system.

It is the successful exploitation of microelectronics by dynamic young entrepreneurs, combined with imaginative software development that has enabled so many attractive electronic energy management products to be placed on the market in recent years. User interest has been exceptionally high, stimulated in part by the glamour of computers and by the growing public awareness that has occurred with the advent of cheap home machines. However, as we have seen from Chapter 4, it can sometimes be rather difficult for someone who is responsible for energy costs to make full and proper use of the technology. Furthermore, not all of the present manufacturers are likely to survive in what is becoming an increasingly crowded and competitive marketplace.

Categories of Supplier

Abba Consultants in their recent book (Computer Based Energy Management in Buildings) identify around 80 firms claiming to be suppliers of electronic energy management systems. Since that publication, the number has probably increased to nearer 100. All types and sizes of firms are included and they each deal with different aspects of the technology. The large number of suppliers has added an extra difficulty to those faced by typical prospective users of the technology; to whom should one go for a particular requirement? Are all the suppliers equally reputable and capable?

Clearly, individual firms have their particular strengths and weaknesses, and in assessing these it might be useful for the user to consider the main categories into which they fall; these are very broadly based, and the characteristics of some suppliers may straddle more than one category. These are summarised below. It should be borne in mind that the descriptions are based on past performance; most firms (particularly the newer ones) are continually gaining experience and improving their capabilities. Most will be in a better position to service future customers' requirements than they have been in the past. Nevertheless, a strong need remains for careful assessments by the customer.

Companies with Previous Experience in Conventional Building Services Control Who Have Now Embraced Computer or Microprocessor-Based Technology

Such firms are likely to have considerable experience of buildings and building services, but may not be necessarily as innovative and imaginative on the computer aspects, particularly with regard to software. However, they are likely to be reliable on installation and commissioning and so are particularly suitable where large and complex projects are involved. Examples of companies who fall into this general category are Honeywell Controls, Landis & Gyr, Satchwell Controls and Johnson Controls. The depth of experience of this type of company in buildings control may be illustrated by the fact that the bi-metallic room thermostat was invented by Dr Johnson of Milwaukee, USA in the 1880s!

Firms with Previous Experience of Computer- or Microprocessor-Based Controls Outside the Buildings Field Who Have Moved Sideways to Embrace the Buildings Market

This type of firm is often able to offer soundly-based, well-developed and reliable products. However, their knowledge of buildings science may not be so complete, and this might show up in the quality and thoroughness of some of the submissions they make to prospective customers. It may be more difficult for them to carry out site surveys to the required depth, to meet the particular requirements of the customer, or to identify aspects of the proposed project that may possibly have been overlooked.

Companies Who Manufacture Very Little But Are Able to Assemble Systems from a Variety of Off-The-Shelf Components Obtained from Several Different Suppliers

Such firms often have an in-depth understanding of computer technology and of the equipment supply situation. Their principal contribution to the systems offered will be software, and this type of firm is often particularly expert in this regard. Their systems are often imaginatively designed but this type of company might potentially be vulnerable to hardware changes by their much larger and more powerful upstream suppliers. Sometimes these changes necessitate software modifications, which could inject delays in the implementation of customers' projects.

Small Firms Recently Established by Innovative Young Electronics Engineers Who Have Identified the Market Potential of Microprocessor Technology for Energy Conservation and Who Are Eager to Exploit It

This type of firm tends to be particularly vigorous and dynamic, partly because of the strong entrepreneurial natures of their founders, but also because of the urgent need to become established and to survive commercially and financially. Naturally, most of the firms in this category are very small (perhaps under 10 employees), and are not well placed to handle large numbers of contracts; sometimes, they make a good living for several years from supplying the needs of one single large customer. Other firms, not so fortunate, mount aggressive marketing and promotion campaigns in order to become noticed. In many cases shortage of investment cash inhibits growth, and these get swallowed up by larger concerns seeking to diversify into what seems a promising new high technology area. Others are keen to remain small so that the founders retain their independence and control over their company's affairs. This type of firm is often interesting to work with. However, the prospective customer does need to assess carefully the actual capabilities of the company and its products.

Examples of Equipment Supplier

Appendix I lists names and addresses of most of the equipment suppliers who have been involved with the ECDPS, either as suppliers to individual demonstration projects or as participants in publicity events organised in connection with the scheme. The list should be used with caution, because this type of information rapidly gets out of date as firms move or change their name. Also, inclusion in the list does not imply any recommendation by the author of any one manufacturer or its products. A more complete list may be found in the Abba Consultants book referred to previously.

Nationalities of Suppliers

As we have seen previously, UK firms did not pioneer the development of electronic energy management techniques; Swiss and US manufacturers were first in the field with large computer-based building automation systems for large buildings and sites, and continue to dominate that end of the market. British firms have made their principal impact at the lower end of the market, with cheaper microprocessor-based systems aimed at small-to-medium sized buildings. Indeed, it is fair to claim that the UK is the world leader in this sector. Furthermore, UK firms are in the forefront of the development of certain types of dedicated control system, lighting control being a prominent example.

However, it is not always possible to assign a definite nationality to particular suppliers, and prospective users who have a firm 'buy British' policy (e.g. local and central government) may have difficulty in satisfying their organisations' criteria in some cases. Suppose, for sake of argument, one defines a 'British' energy management system as one that is conceived, designed and manufactured in the UK by a UK-owned firm; the Transmitton Micropower series of machines appears strongly British against these criteria, but their systems contain many imported microelectronic components. Other 'British' systems are based on imported personal computers (e.g. from Hewlett Packard and Epson). On the other hand, certain 'foreign' firms, such as ISS Clorius, do much of their electronics design and assembly work and software development in the UK. Other foreign companies, such as Honeywell, import entire energy management systems into the UK, but have substantial manufacturing facilities for other products and are major employers of UK labour. Each case, therefore, needs to be considered on its own merits if product nationality is an important factor in the equipment selection process.

Marketing Efforts

The vigour of the energy management system suppliers has been demonstrated by the widespread marketing activities that many of them have embarked upon. Conferences, seminars and exhibitions on this topic have taken place with increasing frequency in recent years, and there is rarely a shortage of representatives from the industry ready and eager to participate. The Department of Energy itself has organised a series of about a dozen events around the country to promote the results and lessons arising from the demonstration projects described in this book. These events included exhibitions by manufacturers, who have reported significant commercial benefit from their involvement. Articles by manufacturers and users of electronic energy management technology appear regularly in the technical press; these have proved extremely useful to other prospective users, but sometimes include over-enthusiastic claims for energy savings that cannot always be easily substantiated (as we have seen in the previous chapter).

The prominent position of UK suppliers has opened up substantial export opportunities which a few companies are already exploiting. JEL, for example, have won a major order to supply a remote monitoring network to Italy. Transmitton have even broken into the difficult US market with an order for a system from the West Coast, not far from the birthplace of microelectronics. However, the majority of the UK firms do not have sufficient marketing resources to address the export market and are understandably concentrating on establishing themselves in the home market first. In any case, the European and overseas market has not been exposed to this technology to the same degree as the UK, and is not yet always so receptive to it; many building operators are probably still unaware that such a thing as an energy management system exists.

The Market Potential for Energy Management Technology

Detailed analysis of the market potential for improved controls for buildings is not presently available; the Building Services Research and Information Association has carried out a multi-

Table 5.1: Estimated total likely sales of energy management systems by the year 2000.

Building Sector (1984 Prices)	*Likely No Of Sales*	*Total Likely Sales Value £ million*
Industrial sites	680	18.2
Office buildings	550	19.3
Hospitals	40	14.0
Government establishments	15	8.4
Retail chain stores	28	11.2
Hotels	20	3.8
Schools	125	17.5
Tenanted buildings (offices and factories)	100	14.0
Total	1,558	106.4

client study of the entire UK controls market, but the results are only available to the participants of that study.

ETSU has conducted a simple analysis of the market for comprehensive energy management systems in small-to-medium sized buildings in sectors covered by the ECDPS (Table 5.1). The figures relate to distributed intelligence networks only and these may in some cases involve large numbers of individual buildings. Also, they omit the sales potentials relating to dedicated control systems and to large building automation systems. The likely number of sales assume a realistic market penetration of 10-20 per cent; this takes into account the barriers to investment noted in Chapter 1.

Chapter 6

HOW TO APPROACH AN ENERGY MANAGEMENT SYSTEM PROJECT

There are a number of logical steps towards successful exploitation of electronic energy management technology, namely:

—preliminary surveys and assessments

—writing a good specification

—selecting suitable prospective equipment suppliers

—assessing tenders and setting up contracts

—staff education and training

—installation and commissioning

—acceptance testing

—utilisation of the system.

The previous chapters have shown that users can encounter problems and pitfalls at each of these stages; but all the stages need to be satisfactorily completed if the project is to succeed. Care and thoroughness is therefore needed throughout.

The remainder of this book is intended to guide the user through the main steps of a typical energy management system project and to provide some of the practical lessons and advice that have emerged from demonstration projects. This should help them to avoid some of the potential problems and to overcome those that do arise. We start in this chapter with preliminary surveys and assessments.

First Things First

Before he commits himself to anything, a prospective user of an electronic energy management system should ask himself whether the technology is appropriate for his organisation; other measures could be better suited to his circumstances. Certainly, it is my firm view that any organisation starting out on an energy efficiency improvement campaign should not do so by primarily investing in sophisticated and demanding measures like energy management systems. Rather, opportunities involving no capital investment at all should be considered. Examples could include:

–administration and policy adjustments (allocating responsibility and accountability for energy costs, appointing an energy manager, etc)

–improving standards of operation and maintenance (getting boilers well maintained and running at maximum efficiency under existing arrangements, making sure that all existing controls are properly working and set up, checking that all HVAC plant including valves, pumps, fans and filters, are operating correctly)

–staff education and training (using poster campaigns backed up with presentations to improve people's levels of energy awareness and good housekeeping).

Once these non-capital expenditure opportunities have been exploited, it will be appropriate to consider all the familiar well-established, low-cost measures that are likely to produce attractive rates of financial return. These might include simple time switches, draughtproofing and insulation and high efficiency lighting.

Only when energy efficiency has been brought to the highest level possible with such means is the organisation most likely to obtain the maximum long term benefits from an energy management system. This is because the system will be monitoring and controlling plant that is well maintained and operating correctly; for example, it is no use trying to connect an energy management system to a jammed heat distribution valve. Moreover, the depth and breadth of information about the site, the plant, the way it functions and the patterns of energy consumptions are likely to be much greater as a result of these preliminary measures. As a result, the decisions concerning purchase of an energy management system are likely to be much more soundly based than they would otherwise have been.

The counter argument to the above is based on the fact that an energy management system itself is a powerful provider of information about the energy performance of a site. Consequently, there may be benefit in purchasing such a system from the outset, and using it to identify what other efficiency improvement measures are appropriate. While there is some merit in this argument, the successful exploitation of energy management technology in this way demands considerable skill and judgement on the part of the prospective user, and carries much greater financial and technical risk.

Carry Out a Good Initial Survey

A decision on investment in energy management technology cannot be made properly unless it is based on a complete set of information. In particular, there is a need to identify and quantify as accurately as possible the likely benefits arising from the system, both directly through improved control and indirectly through the way the system is used (e.g. to identify other improvement measures or to make more efficient use of manpower).

This requires a detailed and comprehensive initial survey or feasibility study. Unless such a survey is commissioned and properly carried out, it will not be possible to establish the potential for reducing energy costs. Nor will it be possible to identify the principal requirements for the system or to calculate the likely rate of financial return from the investment. Such a survey is therefore essential.

Points covered in the survey might therefore include:

–present and past monthly energy consumptions, broken down according to fuel (oil, gas, coal, electricity)

- present energy consumption broken down according to end use (space heating, domestic hot water, air conditioning/ventilation, lighting, industrial processes, etc), together with likely future trends
- present methods and procedures for energy monitoring and control
- present standards of operation and maintenance, good housekeeping and energy awareness
- present levels of satisfaction of building occupants with comfort conditions
- opportunities for improved time/type of day control and the likely savings obtainable therefrom
- opportunities for and benefits obtainable from better set point control of temperatures and humidities
- opportunities for and benefits obtainable from specific optimisation routines (e.g. heating start/stop, enthalpy economisation, oxygen trim)
- opportunities for and benefits obtainable from improved monitoring (including remote monitoring) and the consequent provision of enhanced energy management information and utilisation of manpower resources
- opportunities for and benefits obtainable from electricity cost reduction through demand limiting
- the likely manning and training requirements for operators of the system
- the total overall cost savings benefits likely to be obtained from the system (from energy and non-energy savings), the probable operating costs and the consequent limitation on capital expenditure if an acceptable return on the investment is to be achieved.

One of the results that might emerge from this study is that investment in an electronic energy management system is not appropriate at this stage. This could be because the scope for savings is too limited to justify the probable system costs. Alternatively, it might reveal the fact that the condition of existing plant needs to be improved before connection to an advanced control system can be considered. Furthermore, it may show that the system will probably not be exploited to the full, perhaps because operating staff of sufficient numbers and/or calibre are not available or because building occupants may be alienated; the prospective user must be quite sure he is capable of making full and proper use of the equipment.

Making Out the Financial Case

The results of the survey should allow a reasonably rigorous calculation of the likely financial return on the proposed investment to be made. One uncertainty may be the final equipment capital/installation costs, which await the outcome of a detailed specification and tendering exercise, but budget estimates should be obtainable at this stage. Alternatively, it might be more appropriate to use the financial appraisal at this point to calculate upper limits on justifiable capital expenditure (as a function of the required rate of return). In either case, the results will remain subject to confirmation or amendment when the project details are finalised.

Whatever approach is adopted at this stage, it is particularly important to involve the financial people at an early stage and to present the proposal to the final decision makers in the most appropriate light. Many potentially good energy management system investment proposals founder at this point because the true rate of return on the investment is not revealed or the proposal is not presented in a form which financial accountants can fully appreciate. On the other hand, some projects have gone ahead that probably should never have been approved.

For example, it may not be adequate to describe the likely rate of financial return merely in terms of a crude payback time. This is because a simple payback assumes a well-defined date at which the equipment is installed, commissioned, paid for and begins to save energy; this is never the case in practice, and the implementation and associated costs of the project are likely to be spread over a year or more. Nor are the potential non-energy savings benefits or the likely operating and maintenance costs of the system often fully taken into account.

It is much more meaningful to have a rigorous financial analysis (e.g. discounted cash flow or internal rate of return) which accounts for all factors, including the actual likely timescales of expenditure and utilisation, maintenance and operating costs (e.g. manpower, telephone charges for remote systems, etc), depreciation, non-energy savings and likely future inflation. The implications of company taxation and equipment leasing can also be considered if appropriate.

A project that looks a poor bet on a simple payback basis is unlikely to have its attractions radically transformed through a DCF calculation. However, rigorous financial analysis will provide a more appropriate basis for a Board of Directors to assess the relative merits of a number of investment opportunities that may be competing for a limited amount of available capital. Readers who are unfamiliar with these financial appraisal techniques might benefit from studying an article by J Healey in the July/August 1983 issue of the magazine 'Energy Manager'.

Seeking Independent Advice

Even at this preliminary stage of an energy management system project, it is clear that severe demands may be placed on the prospective user. While equipment suppliers are often willing to carry out initial surveys at no expense to the customer, it is highly dangerous to rely entirely on the figures they supply. For one thing they cannot be unbiased and for another it is unlikely that they could justify the time and effort required to do the job sufficiently thoroughly to form the basis of a major investment decision. Nevertheless, they have a contribution to make and their advice and comments can be of real value.

If the prospective user does not have sufficient or adequate resources to carry out studies of this depth and breadth himself, then serious consideration must be given to using specialist independent consultants. Financial support for this might be forthcoming under the Department of Energy's Extended Energy Survey Scheme, providing it forms part of a wider study embracing all other appropriate energy efficiency improvement measures, such as insulation and heat recovery. The user then has the additional problem of identifying and selecting a suitable consulting firm and meeting the consequent costs; the number of appropriate firms remains fairly small at present, but is steadily growing. It includes small specialist consultancies such as Abba Consultants (Automation), research associations like BSRIA and ERA Technology, and established engineering consultancies such as W S Atkins, Ewbank Preece, Ove Arup, PA Management Consultants, and National Industrial Fuel Efficiency Service. Do not forget to include the costs of these services when calculating the expected financial return from the project or applying for assistance under (say) the Regional Development Grant Scheme.

The value of independent expert advice to the uncertain user can usefully be extended, if required, to subsequent stages of the project, i.e. to equipment specification, tender assessment, installation, commissioning and utilisation. Several of the larger projects described in Chapter 4 have been helped in this way. However, prospective users should be warned not to expect too much from consultants. It will not relieve users of all decision-making burdens, as the consultant will still need to be given a comprehensive and clear initial brief and subsequent supervision. Regular progress meetings will be essential; what one gets out of a consultant depends to a large extent upon what one puts in. Imaginative use of them should, however, enable you to exploit the technical expertise and skills which you may not have in-house, whilst at the same time providing you with an opportunity to contribute your own particular talents (for example, in management, judgement in making decisions and knowledge of your own business and buildings).

Chapter 7

WRITING A GOOD SPECIFICATION

Identifying the Monitoring and Control Requirements

It is important for prospective users to remember that the onus is on them to spell out clearly what they require from an energy management system. Equipment suppliers can help here but users should not depend entirely on their guidance. Ideally, the results of the survey discussed in the previous chapter should provide all the necessary information to enable the monitoring and control requirements to be specified in detail. It should certainly aim to identify which of the principal control options are appropriate. Heating control, for example, is likely to be justified in most applications, but electrical load management would only be sensible if there were loads which can be cycled or shed to limit demand.

The results of the survey should then be able to take the user further and determine what is required from any particular control option. For example, with lighting control it should be possible to decide whether automatic off-switching combined with manual reset is sufficient, or whether automatic on-switching is also required. Furthermore, it should indicate whether options such as photocell control or mains-borne signalling are needed to accommodate the varying lighting requirements of the workforce.

Similarly, for a comprehensive energy management system, the prospective user will need to decide the level and detail of energy management information that is needed. Space temperature and plant status monitoring is on most users' lists of monitoring requirements, but relatively few specifications to date have included a substantial degree of energy consumption monitoring (a prominent exception being the Grampian Regional Council schools project described in Chapter 4). It can be argued that no energy management system is worthy of its name unless it provides sufficient information to allow a wide range of energy management decisions to be made. Many of these decisions require a knowledge of the amounts and patterns of energy consumption, particularly when the case for future additional energy efficiency improvement measures is being assessed. Consequently, remote interrogation of fuel meters should be seriously considered as one of the major monitoring features.

There are other important monitoring options such as occupancy detection (valuable for determining unnecessary heating of buildings as well as for security) and weather condition monitoring (including things like wind speed and direction as well as external air temperature). Unfortunately, many of these monitoring options are expensive and constraints imposed by equipment costs, available capital and required rates of financial return may subsequently rule them out; furthermore, it may not be possible to make full use of the information. Nevertheless, these features should still be considered during the specification stage, and the requirements for them borne in mind when considering future expansions and enhancements of the system.

Identifying Management Information Input/Output Requirements

After deciding on the monitoring and control requirements to be included in the specification, it is necessary to consider the required form of energy management information input and output. With dedicated controls, this may simply concern whether or not to have a chart recorder to log the output from (say) an oxygen trim system, or to provide a digital display and keyboard to facilitate the programming of a lighting control system. However, where a comprehensive system is concerned the quantity and detail of information may require a substantial battery of computer peripherals to enable it to be displayed and interpreted. A line printer will be essential, but it might also be necessary to include plotters or colour VDUs to allow graphs and schematic diagrams to be generated.

Where the quantity of information generated is likely to be especially large, the need for suitable off-line analytical software needs to be considered. This should enable the bulk of analysis and interpretation to be done through the system itself so that operators can concentrate on making energy management decisions. A good example would be the implementation of a targetting and monitoring program, which exploits the remote fuel monitoring that the system can provide; unfortunately, this capability is rarely found outside the well-established large centralised systems, and many of the more recently introduced ones offer no off-line analytical software at all.

Dedicated Controls or a Comprehensive System?

If only a few well defined control options (for example heating or lighting) are involved, then one or more dedicated controllers will probably be most appropriate. However, if a number of functions need to be included, the user will need to decide whether a set of individual dedicated controllers or a comprehensive energy management system is the best way of meeting all or most of the requirements. There are arguments for and against each of these approaches.

Dedicated units will enable the user to shop around and not be tied to any one manufacturer; this may be important, in view of the difficulties and risks associated with selecting suitable suppliers at the present time. He will also be able to proceed at his own pace, perhaps introducing one system at a time and building up experience as he goes along.

However, there are also disadvantages in having an assembly of dedicated controllers rather than a comprehensive energy management system. Firstly, the user will not have the benefits of information input/output at a central station, and may therefore incur significant manpower costs in keeping track of what each of the dedicated controllers is doing. There will always be the serious problem of unauthorised tampering, which can rapidly erode the savings from dedicated controllers. Also, the user will not be able to easily carry out any off-line analysis of energy management data. Furthermore, there is the risk of clashes between control units, (for example, between those for heating and air-conditioning) under certain circumstances. Finally, there is the possibility that a set of dedicated controllers may cost more than a comprehensive system, although this is more likely to be in terms of operation and maintenance than initial capital expenditure.

In making a final choice, the user needs to remember that dedicated units can provide little more than improved control, and are of limited value as an energy management tool in the broader sense. The principal benefit of comprehensive systems is the contribution that they can make to improved standards of energy efficiency in the long term by enabling energy managers to be more effective in their jobs.

Centralised or Distributed Intelligence System?

If the prospective user decides to invest in comprehensive energy management, he will need to consider whether a centralised system or a distributed intelligence system is more appropriate to his needs. The situation is very much one of 'horses for courses'. Both centralised and distributed systems are in principle suitable for single sites; however, a large centralised system may be more appropriate on a large but compact site or on one with complex plant and services, particularly where there is a requirement to get the system in and operating with the minimum of delay and difficulty. On the other hand, a distributed system may be more suitable for a single site comprising a large number of individual buildings (for example a university or government research establishment), where the cost of wiring up a centralised system may be prohibitive or where the need to maintain continuity of control is imperative. Clearly, a distributed system is essential where one is seeking to link large numbers of remotely dispersed buildings into an energy management network through automatic telephone links.

Involvement of Operating Staff

In many cases, the person specifying and selecting an energy management system is not the person (or the only person) who eventually operates it. One of the worst mistakes that can be made is for the project to proceed through the specification stage without consultation with all the people who will actually use the system. As well as professional building services engineers or energy managers, this might include maintenance staff (skilled and unskilled) who have valuable practical knowledge of the plant, its condition and the way it is operated. Consultation with them can prevent simple but typical mistakes, like trying to connect improved controls to plant that is uncontrollable because of major defects or operational requirements. Furthermore, it is essential to obtain the operators' enthusiastic cooperation in order to ensure that the appropriate interaction between man and machine is eventually obtained. It is far easier to create this motivation if they are involved and consulted at an early stage so that, for example, suspicions about possible loss of status or even jobs is not unnecessarily aroused. Staff need to be reassured that management is trying to select the most appropriate means to make more effective use of all resources, not only energy but also manpower; and that this makes the operator's job easier and more interesting by giving him improved tools with which to carry out his tasks.

Need for Restraint

Unfortunately, it is easy to get carried away and include in the specification control functions, monitoring information or items of equipment that are unlikely to be fully used. Only those features which you are confident of being able to exploit fully in the short-to-medium term should be included; it is often possible to expand and enhance the system capability if required at a later date. For example, quite a number of existing projects have included a substantial monitoring capability in order to provide comprehensive energy management information; while the intentions behind this decision were very laudable, in practice it has proved very difficult to exploit this facility because the appropriate off-line analytical software is not available. Other projects have been able to make little or no use of any of the central station facilities because the necessary management organisational adjustments and allocations of responsibilities have not been implemented. In extreme cases, even local dedicated controllers have not been properly exploited because of a lack of motivation and training among the operating staff. Prospective users need to constantly ask themselves the question: 'can my organisation really make full and proper use of the equipment and functions that I have included in my specification?'

Chapter 8

SELECTING SUITABLE PROSPECTIVE EQUIPMENT SUPPLIERS

The Value of Competitive Tender

Once the prospective user has drawn up a well designed specification on the basis of a thorough and comprehensive initial survey, he needs to decide to whom it should be submitted. Under the present circumstances, it is probably a serious mistake to submit it to only one possible supplier. Users should beware of forming cosy relationships with particular firms who may, for example, have supplied conventional buildings controls over long previous periods; although they may provide a competent and effective service in the fields that their experience lies, these firms may not necessarily be the best choice where more advanced technology is concerned.

In my view, there is no satisfactory substitute for a formal tender exercise with all major investments, even where relatively novel technology such as an energy management system is concerned. Firstly, it is the most satisfactory way of ensuring maximum value for money (which may not necessarily correspond to the cheapest system); this is obviously important both in the public sector and in the private sector. Secondly, it can be highly instructive to the prospective purchaser to go through such an exercise, particularly if the submissions received provide several different solutions to the same basic problem; it may even demonstrate that the project is not worth pursuing, at least in its present form. Thirdly, it provides a good insight into the various strengths and weaknesses of the competing companies, and so provides an excellent basis for assessing their true capabilities and the likelihood of their meeting your requirements.

Consequently, my advice is to submit your specification to at least three suitable potential suppliers. However, it is accepted that there will sometimes be sound reasons for only involving a single potential contractor; the project may be very small in capital cost (for example a single dedicated controller), or it may include items in the specification or bespoke requirements that can only be satisfied by one particular firm.

While in most cases a tender exercise will be appropriate, there is still the problem of identifying several suitable firms. The background information given in Chapter 5 should provide some assistance and a number of companies are identified in Appendix I. There are, however, some general points that should be considered.

New or Well-Established Firm?

It will by now be clear to the reader that full exploitation of energy management technology demands a lot from all concerned; this includes the equipment manufacturer, so it is important to bear in mind the likely scale and complexity of a project when considering suitable suppliers. If the project is particularly large, (i.e. including many monitoring and control points, complex building services such as air-conditioning, or special software requirements), it may be appropriate to consider only the larger well-established firms who have the necessary substantial resources to carry through to completion a project on this scale with the minimum risk of delay or technical failure. However, this may mean that the purchaser is constrained to considering mostly foreign companies, and that the systems offered may not necessarily represent the 'state of the art' (the well established larger firms have generally been slower to offer fully distributed intelligence systems than more recent suppliers). While some of the established systems are not as easy to operate as the latest ones, the training facilities offered by the larger firms are often more appropriate to the users' requirements than those of many smaller, more recent entrants. The purchaser therefore needs to make a carefully balanced judgement when selecting a list of suitable contenders for his project if he is to avoid the danger of wasted effort on all sides.

Where smaller projects with less total financial risk are involved, the prospective purchaser can afford to broaden the scope of his tender list. In particular, he should seek to exploit some of the imaginative ideas (in terms of system design, software, performance capability or overall costs) offered by many of the smaller newer firms. However, as we shall see in the next chapter, it is important to separate theory from practical reality when the submissions come in; many smaller firms have been offering facilities that remain to be developed, sometimes on the first hapless customer's premises!

UK or Foreign Supplier?

Many prospective purchasers, particularly in the public sector, are obliged to include only British firms in their list of companies to be invited to tender. However, as we have seen in Chapter 5, it is not easy to define what is meant by a 'British' energy management system. Furthermore, one needs to be careful about discrimination against firms based in the EEC because this can, under certain circumstances, contravene Community Law under the Treaty of Rome.

Nevertheless, it is fair to say that it should be possible to satisfy all the likely requirements of systems for small-to-medium sized buildings using UK equipment. Indeed, UK firms rightly claim to occupy a leading position in this part of the market, at least in Europe.

However, it is not so easy to be so restrictive when one is dealing with a larger, more complex project, particularly where a large single site is concerned. The foreign firms, who comprise most of the larger, more well-established suppliers, continue to dominate this sector of the market. Consequently, a prospective user who has a very demanding specification may find it difficult to exclude such firms from his tender list.

Chapter 9

ASSESSING TENDERS AND SETTING UP CONTRACTS

The Need to Assess Capabilities of Firms

We should now have reached the stage where the specification has been submitted to at least three carefully selected prospective suppliers. Most will wish to visit the site to carry out surveys and to discuss the prospective customer's requirements with him. These interchanges can be very instructive, and all reasonable facilities should be provided so that the suppliers can carry out these preliminary tasks. It may well be that one result of the discussions is that amendments to the specification are suggested, for example if it becomes clear that parts of the specification are unlikely to be satisfied at a reasonable cost. The customer is well advised to take note of what the equipment suppliers suggest and to benefit from their knowledge and experience. Although he needs to rely on his own judgement when considering the final form of the specification, he should be prepared to give way where it is clearly appropriate to do so.

Some of the firms may decline to submit a bid, perhaps because the specification exceeds the capabilities of their organisations or the performance of their equipment. However, this should only happen rarely if adequate care has been given to the selection of potential contractors (Chapter 8). The next task is to assess the submissions received in considerable depth in order to ensure that we achieve in our final selection maximum value for money, whilst at the same time minimising the risk of subsequent technical difficulties.

The first stage of initial assessment is relatively straightforward and obvious, namely to compare the costs of the various bids. Nevertheless, it is worth bearing in mind that the total costs may break down in quite different ways. For example, the balance between hardware, software and installation may vary considerably. Maintenance costs may also differ; normally, this is offered as a separate contract at an annual cost determined as a percentage of the equipment capital/installation costs (typically between five and fifteen per cent). Training costs may also vary.

Some of the bids may be immediately eliminated because the total costs are substantially above the average level or the financial budget set for the project. Higher than average costs may reflect a reluctance to take on the task you have offered.

A smaller group of bids may then remain with total costs grouped around a level well below the initial average level and within the budget limits. The task of sorting out the one representing best value for money will involve the application of management skills and technical judgement. In this respect, the assessment has many features in common with other investment decisions that most prospective purchasers will have made. However, where energy management systems are concerned, attention to the following points may prove of particular value.

Have All Likely Costs Been Included?

Go through the submissions very carefully to determine how closely they correspond to the agreed specification. For example, make sure that full account has been taken of all likely training needs. Also, check that installation costs correspond to a full 'turn-key' operation, if that is what you are seeking, and that the calculations do not assume that substantial wiring or alteration work will be supplied by the customer.

Is the Proposed Equipment Fully Developed?

As mentioned previously, it is extremely important to ascertain the exact stage of development of all hardware and software being offered. Beware of falling into the trap of having your project used as a development testbed, because everything is then likely to take far longer than expected; 'a small matter of software development' can translate into many months of delay and frustration. It is good practice to visit the leading contenders and to interrogate them closely on this point.

Ask them the simple question: 'can your equipment presently do all that I am asking in my specification?'. Make sure you get a straightforward answer, and if it is negative establish clearly what further development work is required and the degree of effort and complexity involved in completing it. Press for a realistic assessment of the timescale for completing such work, and remain sceptical about stated timescales; these are rarely adhered to in practice, because of a natural tendency of firms to underestimate the effort, time and cost of development work.

Where it is stated that equipment is already developed, ask to see a demonstration on the firm's premises or to have the name and address of a recently satisfied customer with a similar system.

Can the Firm Really Tackle the Task?

Another area of potential difficulty is the availability of resources in the company to execute your project, if they do receive your order. Many of the suppliers are small and so are naturally anxious to fill their order books to maintain a reasonable cash flow. In many cases, expansion is difficult because of a shortage of available investment cash, particularly if existing borrowings are substantial. Consequently, some firms may already be near the limit of their ability to service new orders satisfactorily and within a reasonable timescale. It is a good idea, therefore, to question the representatives of the firm closely on this point; you do not want to see your project slow down or stop because all resources have been allocated to other customers.

Will the Necessary Expertise be Available?

Because many of the equipment suppliers are small, a substantial proportion of the expertise available in the firm often resides with just one or two prominent individuals. These days such people are highly mobile, and may suddenly move to a more attractive position with a new employer. You should examine the likely consequences of this happening, and satisfy yourself that your project would not be put at excessive risk if it did. This boils down to making judgements about all the principal people likely to be involved in your project, and illustrates the value of meeting them on their own ground at their firm's premises.

Obtaining Final Financial Approval

At this stage, it may well be that the customer only has approval in principle for the funding of his energy management system up to an agreed budgetary limit. If the final costs of the selected system are higher than this limit, he will need to justify the additional expenditure when seeking final financial approval. The difference between the initial estimate and the final costs following the tender exercise will to some extent reflect the prospective user's knowledge and experience of the technology; if the difference is too great, there may be no alternative but to modify the specification or scale of the project, or to abandon it altogether. On the other hand, most funding bodies are willing to consider and approve modest increases in budgetary limits, providing the reasons for the increase are understood and justifiable, and that the investment remains a good one. It is important, therefore, to document the financial circumstances of your project carefully so that a well argued case may be articulated for final financial approval.

Setting up Contracts

Assuming all goes well and the project receives funding approval in its final form, it is vital that the actual order is placed by means of a properly drawn up contract with a comprehensive programme of work based on the agreed final specification (it may be possible to use standard sets of contract words developed by advisory bodies such as the IEEE). This may sound like obvious common sense; however, the author is aware of a number of projects where considerable difficulties have arisen through failure to establish an appropriate contractual relationship. For example, in at least two cases the contract was signed first and the specification drawn up afterwards; the inevitable result was dispute over what exactly was or was not included in the contract for the agreed price. All parties to the project need to understand exactly what is expected of them, including the customer; he may be responsible for items such as local wiring and installation work and liaison with utilities for the supply of metering and telephone connections. Many of the demonstration projects with which the author has been involved have suffered difficulties arising from confusion over who was responsible for doing what and by what date.

Realistic Timescales for Installation/Commissioning

One of the most important features of the programme of work should be the agreed timescale for implementation of the project. This should include well-defined target dates for key milestones, such as installation of hardware, loading of software, commissioning of the system, and handover to the client. There is a danger of customers demanding and suppliers agreeing to quite unrealistic schedules for these dates, often simply to meet a forthcoming heating season; the likely result is frustration and disappointment as the slippage begins to build up. A major energy management installation cannot be installed and set to work in just a few months; most of the projects described in Chapter 4 have taken over a year to implement. The customer needs to have frank discussions on this point with the selected supplier, so that a reasonable timescale can be agreed. This means that it will not always be possible to start making substantial savings in the next heating season, particularly when one remembers the learning process the user must go through before he can fully exploit the system. This fact should be reflected in the user's financial appraisal of the project as well as in the contract.

Progress Payments

In practice, once the project commences the only effective leverage that the customer will have over the equipment supplier is payment for a job properly executed and the prospect of future repeat orders. It is important to preserve the former as long as is reasonably possible, and it is therefore a mistake to include in the contract terms provision for progress payments that may be premature. Some existing users have suffered considerably through paying the bulk of the contract price on delivery and physical installation of hardware. They have then found themselves in proud possession of valuable equipment that cannot be used because the software that allows it to function remains to be delivered (usually because it has still to be developed!). The customer finds it difficult to bring effective pressure to bear because he has already parted with most or all of the money. The contract should spell out clear milestones for appropriate progress payments, for example:

- delivery of hardware
- installation of hardware
- implementation of local (e.g. outstation) software
- implementation of central station software
- implementation of remote communications links
- handover to customer (following acceptance testing).

The last payment would normally correspond to around 10 per cent of the total contract price, whilst the previous payments should reflect the relative values of the work carried out to date.

Chapter 10

STAFF EDUCATION AND TRAINING

Motivation and Training of Operating Staff

In many energy management system projects, the person responsible for the technical and financial decisions is not the person or only person who will be involved in the day-to-day operation of the system. For example, hourly-paid operations and maintenance staff may need access to the data or controls at the central station in order to make routine adjustments and take records. Only more substantial changes and analyses may be carried out by higher level staff such as the energy manager or site engineer. These different levels of access to the system can be controlled by various entry passwords corresponding to the various levels of management responsibility.

However, the important point here is that everyone involved with the utilisation of the system understands what it is intended to do and how it should be operated, and is motivated to make the best possible use of the new equipment. This ideal state of affairs can only be achieved if the education and motivation process starts well before the system is installed.

As mentioned in previous chapters, it is good practice to inform and consult all the staff likely to be involved in the operation of the system at the time that the specification is being determined. Only then can the project benefit from the practical knowledge and experience that may be available, as well as from the enhanced likelihood that the operating staff will be motivated to make full and proper use of the equipment. Presenting it to them as a 'fait accompli' and simply asking them to get on with it is likely to trigger off a serious 'Luddite' reaction, particularly if there are fears of possible job losses resulting from the introduction of the new technology.

The main message to all the staff involved in the project is that it is intended to make them more effective in their jobs, thus making their lives easier; at the same time it will reduce the operating costs of the organisation and so help to secure the long-term employment prospects of all its staff.

In some cases, it will not be possible to give such assurances, because reductions in manpower may be intended to be one of the principal sources of financial return from the investment, once it has been installed and proven. Under these circumstances, it may prove to be extremely difficult to implement the system successfully, particularly if the operating staff see the reward for cooperation as probable redundancy. The large boilerhouse automation project at Bowater's paper mills described in Chapter 3 has been less than fully successful, partly as a result of problems of this nature. Such difficulties are likely to beset the introduction of new advanced labour-efficient technology in older, hard-pressed industries.

Once all operating staff have become reasonably motivated to make good use of the new technology, the next task is to ensure that they have sufficient training to actually succeed in doing so. The levels of training that are needed will, of course, vary according to the levels of responsibility that are allocated.

Manual workers can often best be trained 'on the job' by allowing them periods of working alongside the equipment supplier's technical staff during the installation and commissioning phases. If this approach is adopted, it is vital to allow sufficient time off normal routine duties so that all the necessary information and instructions can be assimilated.

Higher level operating staff may need short full-time intensive courses. The larger equipment suppliers, such as Johnson Controls, have proper training schools set up specifically for this purpose at their headquarters, and many firms have a fully operating system set up for demonstration purposes in their offices; this can be made available for customers' staff to practice and become proficient on.

Whatever approach is adopted, the training requirements should be incorporated in the final equipment specification and the direct and indirect costs included in the total project costs. The training facilities provided by the supplier should be one of the important factors taken into account when making a final investment decision. In due course, further supplementary courses may be needed after the system has been implemented, particularly if the scope of its operation needs to be expanded. There are few of these available at present, but a limited number are run by the Buildings Research Establishment at Garston.

It is likely that one of the main training problems will stem from the changes and turnover of operating staff that will inevitably occur during the lifetime of the system. Some of the expertise can, of course, be transferred during hand-over periods. Nevertheless, standards of system operation are likely to fall unless the equipment is relatively straightforward to use and is well documented by clearly-written instruction manuals.

The Ladbrokes Hotels' distributed intelligence remote energy management network project described in Chapter 4 has particularly suffered from the high staff turnover in the hotel industry. Many of the maintenance men that were trained by JEL during the installation and commissioning phases are no longer in their original posts. This would not have mattered so much if the operation of the JEL outstations had proved simpler and easier to understand than has actually been the case. Some of the potential benefits of the system have not been realised because the incoming replacement staff have not been able to grasp quickly what the equipment does and how it should be operated. Unfortunately, it is a common human tendency only to refer to the instruction manual when all else has failed!

Informing Building Occupants

It is important not to lose sight of the fact that energy management technology for buildings is used to modify the comfort conditions of occupants. In some cases, the result of installing an energy management system may be to lower standards of comfort; it is little consolation that space temperatures and/or lighting levels might have been excessive before. It is therefore extremely important to inform all occupants of the building fully about what is being done and why; otherwise the consequent adverse reactions can be very severe, and in the worst cases this may even jeopardise the future success of the project, perhaps even causing it to be abandoned.

Problems of this nature are particularly likely to occur if occupants already consider (rightly or wrongly) that their comfort conditions are inadequate, even before the new system is installed. As we have seen, projects in hospitals and schools have suffered particularly badly in this respect.

It is therefore good practice to inform the occupants at an appropriate early stage of the project, possibly soon after a firm decision has been made to proceed. They should be given fairly comprehensive details of what the system is, how it will operate, what the benefits are hoped to be, and (most importantly) how individuals should interact with it through, for example, lighting reset switches. It is important to ensure that this information disseminates to all members of staff; this means that a suitable notice may need to be sent to each and every individual, because experience suggests that informing just supervisory staff rarely succeeds in fully promulgating the intended message. It is also a good idea to follow up the initial notification with presentations to selected groups of people, particularly union representatives and supervisors, so that clarification can be provided, objections aired, and confirmation obtained that occupants actually do understand what is intended. Furthermore, these meetings may yield additional practical information and advice which could prove valuable contributions to the eventual success of the project.

This close consultation is particularly important with network energy management projects, where staff members may be rather isolated from headquarters staff and be naturally hostile to measures that are being imposed upon them; schools are a good case in point.

Chapter 11

INSTALLATION, COMMISSIONING AND ACCEPTANCE TESTING

Time Slippage in the Agreed Schedule

It has been noted in earlier chapters that many existing energy management system projects have been set up on the basis of quite unrealistic timescales for the implementation of the various phases. The user is understandably keen to start reaping benefits from his system and, in particular, to begin making savings during the next heating season. Indeed, financial approval may have been obtained on the basis of a cash-flow analysis that assumes these early savings. Equipment suppliers, naturally anxious to please and impress, often agree to schedules which cannot be maintained in practice. Slippage, with the attendant deterioration in the relationship between customer and contractor, is then almost inevitable. It is hoped that future users will avoid the worst of such problems by following the advice given in Chapter 9 and by incorporating more realistic timescales in the contractual programme of work agreed with the supplier, taking into account all known circumstances and likely difficulties.

However, despite the best efforts of all parties involved in the project, it is likely that many problems (some of them unforeseeable) and consequent delays will still occur. This is not necessarily a reflection on the competence of the equipment suppliers, but rather on the fact that energy management technology for buildings remains relatively novel; most parties concerned with its development and implementation are still on a steeply rising learning curve, which places severe demands on them. It is important that the user understands why and how problems can occur so that the most appropriate remedial actions can be determined and adjustments made, if necessary, to other aspects of the project.

In this context, it may be useful to consider some of the typical difficulties experienced in the installation and commissioning of systems being supported under the ECDPS.

Problems with Existing Plant

A principal source of problems is likely to be the customer's own plant and services. An energy management system needs to be interfaced with existing plantroom equipment, and during the course of installation many defects that have lain hidden for years are likely to be unearthed; they will need rectification before work can proceed. Mixing valves, for example, are often found to be beyond their useful lives and in need of replacement.

This illustrates the wisdom of only installing an energy management system against a background of high standards of operation and maintenance. The preliminary survey discussed in Chapter 6 may identify items needing repair and maintenance, and these should be attended to before embarking on the main project. There is little point in installing advanced control systems if the plant it is connected to cannot in fact be controlled because it is not operating correctly.

Software Development Problems

In previous chapters, considerable emphasis has been placed on the need to establish the exact state of development of the systems that are being considered and to select one that already meets the customer's specification as closely as possible. However, some development work may still be necessary to meet particular requirements; this is likely to involve software rather than hardware and has frequently proved a source of much delay. Equipment suppliers tend to grossly underestimate the time and effort required to complete work of this nature, in particular to 'de-bug' an otherwise finished program. Users should try to minimise the likelihood of this happening by avoiding inclusion of items in the specification that cannot be satisfied by existing system facilities.

However, nothing in this field stands still for long and new products are constantly being released. Consequently, even when the above advice is followed the customer is likely to come under pressure from the supplier during the course of installation and commissioning to adopt his new improved operating software system that is currently 'under development' rather than use the 'inferior' existing system. Offers of this nature are clearly tempting, since they often hold the prospect of greatly enhanced performance. However, before accepting them, the user should satisfy himself that he is not likely to fall victim to excessive delays or, even worse, find that much of his existing hardware and software is incompatible with the new system. For example, new outstation software may require changes to central station software as well.

Communications and Data Collection Problems

Many of the remote network distributed intelligence projects supported under the ECDPS have had problems in establishing adequate communications between outstations and the central station. Some of these stemmed from shortcomings with the British Telecom auto-dial/modem units that users were obliged to employ before 1982. These units were difficult for the energy management system manufacturers to understand and interface with their own equipment, largely because the documentation was poor. This problem has been much reduced by the abolition of the BT monopoly in this field, and suitable communications units are now available from specialist suppliers such as Dynamic Logic.

There have also been some problems in obtaining monitoring data through remote energy management networks. For example, information can be seriously corrupted by noise on the electricity cabling of the building; the Sainsbury supermarket project has been prone to difficulties of this sort with the remote reading of electricity sub-meters. Inadequate earthing is a common cause of such problems, and can be particularly difficult to identify and correct. In the meantime, the value of the energy management system to the user may be diminished because the data contains much garbage.

Other projects have encountered problems in maintaining reliable telephone links through the auto-dial units (again possibly through electrical noise), and obtaining satisfactory rates of data transfer; routine monitoring reports have taken far too long to come through to the central station and have generated excessive telephone charges.

The Need for Regular Project Progress Meetings

Because of the technical difficulties and delays that can be encountered with a typical energy management system project, it is very important that regular project progress meetings are held (at least every two months); only in this way can the user obtain a good understanding of the state of progress and the problems being encountered.

The customer should take the chair and control these meetings; but the discussions should be open and frank so that consensus decisions can be reached on how the problems should be overcome and what the forward work programme should be. On the other hand, agreed target dates should mean what they say and parties to the project should be held accountable for any serious slippage that occurs.

Controlling a project in this way can be a rewarding experience and an effective means of achieving the intended aims. Moreover, it should help the supplier to meet his obligations by making it quite clear to him exactly what is expected of him during the periods between meetings.

Behavioural Problems with Occupants

The need to inform occupants of the building fully about the energy management system project has been covered in the previous chapter. If this is not done adequately, there is a danger that reactions to problems of the type described above will be unnecessarily severe. Teething troubles with the system during installation and commissioning are likely to cause temporary abnormalities in comfort conditions; it is annoying to have rooms too cold (or too hot), or lights going out when you need them, but it is tolerable if you know the difficulties are likely to be short-lived and are connected with a worthwhile project. It is quite another matter for the same things to happen for no apparent good reason. If the occupants have not been adequately informed beforehand, on discovering the cause of the difficulties they are likely to associate all future comfort problems (rightly or wrongly) with the energy management system. As the manager of one schools project has remarked 'if a blackboard falls off the wall, they say it's that damn energy management system again!'

Once problems of this nature have been encountered, it is not easy to rectify the bad image that has been generated. The only course of action, apart from eliminating the teething troubles as quickly as possible, is to mount a retrospective information campaign; needless to say, it is far better to address this requirement at an earlier stage of the project.

The Need for a Formal Acceptance Testing Schedule

Let us assume that installation and commissioning of the energy management system has proceeded to an advanced stage. A number of problems have been encountered along the way, and these have been overcome by one means or another. Progress payments have been made as hardware has been installed, and various elements of the system have been brought into operation. Training of all the operating staff has proceeded in parallel with the commissioning process.

However, the formal hand-over of the system to the customer and the final payment on the contract cannot take place until there is a clear agreement that what has been provided by the equipment supplier satisfies the customer's requirements as laid down in the contract. This sounds a simple matter, but is rarely so in practice; it is vital that a formal procedure is set up for carrying out a series of acceptance tests on equipment performance and comparing the results of these tests with the contract programme of work. If this is not done, there is likely to be considerable disagreement over what the system can or cannot do, arising partly from the fact that the two parties tend to speak different technical languages and may not recognize each other's descriptions of particular functions.

Agreeing the Details of the Tests

The best policy is for the supplier and customer to agree a set of well-defined acceptance tests well before the commissioning procedures are complete; the requirement for such tests should also be incorporated in the contract, even though it may not be possible to define the actual procedures at that stage. The tests should cover all the principal monitoring and control functions included in the specification, together with information input and output capabilities, communication procedures and data analysis facilities.

Implementation of the Tests

The tests should be carried out over a pre-defined period (of not more than a couple of weeks) in the presence of suitably qualified representatives of both parties who should each be empowered to certify that individual tests have been completed to their satisfaction. Some tests may not produce a positive result because of intractable underlying circumstances; these should be well documented so that a satisfactory adjustment in the final agreed contract price may be negotiated.

Having the customer's staff working alongside the equipment supplier in this way has considerable training value as well as helping the project to proceed towards a mutually satisfactory conclusion. However, it is likely to be time-consuming and technically demanding, especially with a large and complex system; the customer may feel that sufficient staff time cannot be spared, or that his technical expertise would not enable him to make the sort of technical judgements that the tests demand. Consequently, it may prove beneficial to extend the scope of the brief to an advisory consultant to cover not just the previous stages of equipment specification, selection, installation and commissioning, but final acceptance testing as well.

Chapter 12

MAKING THE BEST USE OF THE SYSTEM

The author has been involved with several energy management projects where the systems have now been fully installed and commissioned; the users are in the process now of learning how to use their equipment in the most effective way. This learning process is proving to be more protracted than expected, and it is likely that it will be several years before most typical users are able to exploit the full potential of their systems.

This illustrates why it is inadequate to express the expected rate of return from the investment merely in terms of a simple payback period. Equipment suppliers often promise that the capital outlay will be returned in less than two years, but it is likely to take many future users longer than that before they are able to use their systems properly!

In the meantime, many lessons are being learned as a result of the experiences gained by the existing users. These can be fed back to the benefit of future users.

An important lesson has been that energy management technology is of limited value in itself; it requires a close interaction with the people responsible for operating it if long-term savings are to be achieved. This technology differs in one major respect from most other energy efficient improvement measures in buildings like insulation or heat recovery, which are capable of saving energy directly and automatically without any such interaction. You cannot walk away from an energy management system like you can from roof or wall insulation and expect it to save energy over a long period.

In addition, there have been other important lessons; some of them will be described briefly in the following sections.

Problems of Handling the Data

With remote distributed intelligence networks, the total number of monitoring points can become very substantial, either because the number of buildings connected to the system is large (e.g. in the Hereford & Worcester schools project described in Chapter 4) or because the level of monitoring detail for each building is very high (e.g. in the Grampian Regional Council schools project). Under these circumstances, the amount of information (on comfort conditions, fuel consumptions, alarm messages, etc) arriving at the central station can reach alarming proportions. It is all being requested for good energy management reasons but is of limited value unless it can be assimilated, analysed and acted upon.

Unfortunately, the quantity of information can quickly become too great to handle properly and the operators may begin to disappear under a mountain of computer output. The situation is likely to become worse as more buildings are added to the system which started out as pilot projects. It is becoming increasingly clear that considerable off-line analytical capability is going to be required if the quantities of data likely to be generated by large-scale energy management systems are to be usefully interpreted. The operators may need to be presented only with filtered information on which they can act directly rather than with everything the system is capable of providing. Thus, while a high level of monitoring capability needs to be retained in most cases, the system itself should sort out those reports reflecting normal acceptable circumstances from those that clearly need some human attention i.e. an energy management decision.

At present, few systems offer much in the way of intelligent analytical softwarc of this nature. Many can produce, for example, useful trend logs of fuel consumptions, but to the author's knowledge none of them yet provide an energy 'management by exception' facility of the type described above. No doubt such features will be developed in due course, as the requirements of the market become clearer and the reservoir of user experience begins to build up. In the meantime, most suppliers are concentrating on developing and improving the operational software of their systems rather than their analytical capabilities.

Reacting to Information Output at the Central Station

In additon to the problem of handling large quantities of information, other barriers to full and effective use of an energy management system have emerged from some of the existing demonstration projects. One of the principal ones has been the failure of the user organisation to make the necessary management and administrative adjustments to enable the information presented at the central station to be acted upon. An example is the Ladbroke Hotels project. There the central station is located in the largest of the 15 hotels included in the project, (the Westmoreland in London), but unfortunately the operator of this station has no authority to act on any of the information received, except where it concerns his own hotel. Moreover, his regular operations and maintenance duties at the Westmoreland allow him little time or effort to even think about the energy management problems of other hotels. Even if they did, he would certainly not be in a position routinely to adjust comfort conditions in any of the other hotels, because authority to do this resides with the local managers who are unlikely to tolerate his interference. As a consequence, the bulk of the facilities provided by the system, particularly those relating to remote monitoring and control, are not being used to the full. As a result, the investment is likely to remain a poor one, at least until the barriers to full exploitation are removed.

The lesson here is that future prospective users must satisfy themselves that their organisation is capable of utilising the technology, not only in terms of technical expertise, but also in terms of available manpower resources and management responsibility. If necessary, additional resources must be allocated and appropriate energy management responsibility and authority given to the operators of the central station; otherwise, the system is likely to prove a waste of money.

Maintenance

Users of energy management systems should not forget that they have many complex electronic, electrical and mechanical parts, all of which need proper regular maintenance if they are to function correctly over long periods. This is particularly important with centralised comprehensive systems which rely heavily on the correct operation of the CPU for most control facilities. Modern electronic components are renowned for their very high reliability, but faults can still occur. It is with electro-mechanical equipment that problems are most likely to arise; disc-drives and printers at the central station and relays in the plantroom are the sorts of component likely to give trouble.

The simplest arrangement for maintenance is through a contract with the equipment supplier, usually based on annual payments of a fixed percentage (around five to ten per cent) of the initial total capital cost. However, users need to consider carefully exactly what is being offered; some contracts only cover calls out to repair faults as and when they occur and not regular preventive maintenance. Where regular maintenance is included, users need to establish some means of determining exactly what has in fact been done and when. Finally, the user needs to consider whether the resources of the supplier are adequate to provide the level of maintenance needed and to take this into account when considering the choice of system. Certainly, the risk of discomfort and upset with building occupants resulting from frequent malfunction of an energy management system needs to be minimised by making the best possible arrangements for future maintenance.

Upgrading the System

Users should be wary of allowing their system to stand still; even when a system has been fully installed, commissioned and utilised, the process of improving its capabilities should continue. This might include purchasing further operational or analytical software packages as and when these become available, either from the original equipment supplier or possibly in the future from specialist software houses. By becoming reasonably proficient in the language used in the system (most commonly BASIC), users themselves may begin to produce their own programs; this can prove a particularly powerful way of upgrading the system, because the user is always in the best possible position to determine his actual requirements.

The user should also continually consider the scope for expanding the system physically, in terms both of the functions it carries out and of the numbers of plantrooms and buildings connected to it. For example, many existing comprehensive energy management systems do not include lighting control, possibly because the user was not sufficiently aware of the cost-effective techniques that might have been available at the time the original specification was drawn up, or because he wished to concentrate on reducing heating costs in the first instance. Once the initial aims have been achieved satisfactorily, a good case may exist for introducing some form of lighting control using the energy management system as a time clock and calendar. Most of the lighting control systems now produced by specialist firms to cover typical building circumstances can be interfaced very easily in this way. Similarly, boiler control units might also be introduced and connected up to the system. However, each potential expansion of the system must be carefully assessed in terms of its technical feasibility and its financial viability.

Chapter 13

FUTURE DEVELOPMENTS IN ENERGY MANAGEMENT TECHNOLOGY

Dedicated Control Systems

A wide range of dedicated controls is now available for buildings; many of these have already been described in Chapter 3. The main future trend in this area is likely to be a steady reduction in cost as exploitation of microprocessor technology proceeds and the benefits of increased production levels begin to come through. At the moment, a 'chicken and egg' situation still exists; mass production awaits the establishment of a larger market, which itself needs the lower unit costs that only mass production can provide before sales can be built up.

In particular, there is considerable scope for low-cost advanced heating control devices; many of the programming and control features now available for around £1,000 in units aimed at commercial, industrial or public buildings will probably soon be incorporated in domestic heating controllers selling for under £100. Similar cost reductions are likely in other areas of dedicated control system development; oxygen trim systems might also come down sufficiently in scale and cost eventually to begin to penetrate even the domestic sector.

However, lack of appropriate user expertise is likely to be a significant factor in limiting the full penetration of some potential markets; many existing users, particularly in the domestic and small commercial buildings sectors, are unable properly to understand, operate and maintain existing types of controls, let alone new advanced types. They are therefore unlikely to become completely receptive to the new technology for some considerable time yet.

Furthermore, as one comes down the building size range the user will often not be the initial customer for the control device; the small user typically wants a 'heating system' with some form of suitable 'controls', and it is likely to be plumbers, heating engineers and builders who will continue to specify the actual nature of the controls in this market. These people tend to be technologically conservative and particularly price-conscious and will probably continue for some considerable time to specify a £20 electromechanical time clock and a room thermostat in preference to a £100 microprocessor-based heating controller. The full transfer of the technology into the small buildings market place is therefore likely to be protracted.

Where uptake of particular dedicated control technologies has previously been restricted by high installation costs, there are likely to be major advances brought about by imaginative application of microprocessor and communications technologies. The obvious example will be lighting control incorporating mains-borne signalling techniques, which will become increasingly refined. The same techniques can also be applied to the remote control of any electrical device, including pumps and power operated valves; so there is no technical reason why mains-borne signalling should not be used on most aspects of building services control, including heating, ventilation and air-conditioning, with similar consequent reductions in installation costs. The principal difficulty will be satisfying prospective users that there is no significant danger of interference with electrical equipment within the building, or with similar control systems installed in neighbouring buildings.

Comprehensive Energy Management Systems

The main development here is also likely to be a continuing reduction in hardware costs, thus extending the cost-effective applicability of the technology further down the buildings size range. The limiting factor will be installation costs, which now can account for up to 50 per cent of total project costs; some reductions can be expected here through full exploitation of distributed processing and improved communications techniques, including mains-borne signalling.

A further development is likely to be an increased dispersal of the central station functions to outstations once outstations can communicate with each other as well as with the central station. This facility should enhance the effectiveness of maintenance engineering staff, who will not need to be at their headquarters offices to communicate with outlying buildings connected into a remote monitoring and control network. Several firms, including Trend Controls and Information Transmission, already offer equipment with this capability.

Further developments are likely to include improved software, and particularly software that can easily be used by people who are not computer experts. We can also expect individual systems to offer wider ranges of operational routines, together with appropriate off-line analytical programs; for example, these will help with the burden of interpreting data or providing targetting and monitoring procedures.

Some of the control functions that are at present exploited through dedicated control units are more likely to be incorporated into comprehensive energy management systems in the future; the most important example will be lighting control, which is still rarely included in existing projects, even though it can generate substantial energy cost savings. Another control technique likely to be found more often in future comprehensive energy management system projects is boiler control, and in particular oxygen trim; to the author's knowledge no system at present includes this measure, even though there is no technical barrier to its incorporation. In view of the importance of maintaining fossil fuel combustion efficiency, we are likely to see it as part of many comprehensive energy management systems in the future.

Sensors and Actuators

Development of the microprocessor-based hardware elements of electronic energy management technology has proceeded apace and is now capable of satisfying all likely user requirements for the foreseeable future. While software is not generally so advanced, it is rapidly being developed to match the capabilities of the hardware and management needs of the user. What is being left behind, to a large extent, is the development of improved sensors and actuators.

In many cases, we are trying to connect 1980s computer equipment to 1950s control hardware; this development gap might reflect the fact that working on the latest microelectronics is far more interesting than trying to improve the performance, accuracy and reliability of valves and sensors.

Using existing sensors, the control accuracy that can be achieved still gives rise to a significant excess fuel consumption. For example, the overall accuracy of space temperature set-point control quoted for some comprehensive energy management systems is only 1°C, which can correspond to an average excess heating consumption of 10 per cent. A more reasonable norm is 0.1°C, but in many cases system accuracy prevents this from being achieved; it is to be hoped that suppliers will actively strive towards such a standard as improved temperature sensing and heat flow control techniques gradually become available.

Air-conditioning control is beset with even more severe problems which arise from the limitations of present humidity measuring techniques, which are based on the very non-linear changes in the electrical characteristics of lithium chloride paste as its moisture content varies. These devices have proved notoriously unreliable, and the resulting control errors are estimated to give rise to as much as 30 per cent excess refrigeration consumption. There is thus an urgent need for a device with superior performance, probably based on semiconductor technology.

Similarly, there is considerable scope for improving the performance of flue gas sensors for boilers, whilst at the same time reducing the size and costs. Semiconductor technology probably holds the key here also and could open up a vast untapped potential in the small boiler size ranges.

Unfortunately, all the developments described are likely to involve some very expensive research and are therefore outside the scope of the small companies who manufacture most energy management equipment. It will probably require the involvement of the major semiconductor companies and strong government support before we can expect to see substantial improvements in this area.

Compatibility of Systems

One of the annoying things about present-day energy management technology is the fact that nearly all the different systems are incompatible with one another, in terms of either the hardware or of the software aspects. Investment in a pilot-scale project using equipment from one manufacturer may confirm the value of the technology but prove unsatisfactory in some way; it is rarely possible to expand that system using equipment from another manufacturer which overcomes the problem. This means that one has then either to abandon the equipment first purchased, or to set up two independent parallel systems. Moreover, it increases the risks associated with a project because one usually becomes tied to a particular supplier, who may give unsatisfactory service or even go out of business. Even if the initial equipment and supplier are satisfactory, rapid developments may make the system purchased obsolete and incompatible with more recent versions.

There is therefore a strong need for the wider adoption of industry standards in this field, and it is to be expected that the various manufacturers will begin to communicate and cooperate more on this topic. Some tentative moves have already been made towards developing a set of British Standards for energy management equipment, and improvements in this area should begin to emerge in the near future.

This leads to the question of compatibility of energy management equipment with other computer systems. Many organisations have central mainframe computers or distributed networks for tasks such as point-of-sale accounting in the retail sector or booking rooms in the hotel industry. These organisations would obviously like to use existing equipment for energy management, or to integrate an energy management system into their other computer networks. The former option is at present impractical, because of the costs of developing the necessary energy management software (this would not normally be available other than from an energy management equipment supplier). Also, not all computer systems are suitable for carrying out closed-loop control functions. Nevertheless, it is to be expected that software houses will recognise a market here and develop this particular business opportunity.

Linking an energy management system into another computer is also difficult or uneconomic in most cases, but several suppliers are beginning to offer inter-system compatibility as one of their products' special features. A good example is the 'Microreporter' from Information Transmission, which opens up opportunities for carrying out some particularly effective targetting and monitoring procedures.

The General Application of the Technology

As time goes by and uptake increases, attitudes towards electronic energy management technology are likely to change gradually. It will begin to be looked upon less as an immediate energy efficiency improvement opportunity and more as a means of improving the overall standard of energy management in the long term. Many of the immediate benefits of an energy management system can be obtained more cheaply with existing conventional controls, but the latest techniques provide a powerful opportunity to ensure that these benefits are maintained or improved upon. They are thus an aid to the effective utilisation of all resources available to an organisation, not only energy but also manpower and materials. Judged merely in terms of reduced energy costs, many comprehensive energy management projects have produced only modest financial returns, but judged in terms of the (actual or potential) contribution that they can make towards creating a better-run organisation, most of the installations should eventually prove worthwhile.

Appendix I

REGISTER OF ESTABLISHED SUPPLIERS

Manufacturer/supplier	*Activity(ies)*
AEG Telefunken Market Street Maidenhead Berks SL6 8AE Tel: (0628) 39171	Centralised comprehensive energy management systems (micro-computer based)
AFA-Minerva Ltd Security House Grosvenor Road Twickenham TW1 4AB Tel: 01-892 4422	Distributed intelligence and micro-computer based centralised energy management systems
Allen-Martin Electronics Ltd Marlborough Works Thompson Avenue Wolverhampton WV2 3NP Tel: (0902) 58942	Dedicated heating control Dedicated lighting control Distributed intelligence and micro-computer based centralised energy management systems
Atlantic Instruments Ltd 1 Minto Avenue Altens Industrial Estate Aberdeen AB1 4JG Tel: (0224) 895024	Distributed intelligence and micro-computer based centralised energy management systems

Manufacturer/supplier	*Activity(ies)*
Building Systems (Marketing) 1 Crossland Road Redhill Surrey RH1 4AN Tel: (0737) 62507	Distributed intelligence and micro-computer based centralised energy management systems
Burgess Energy Systems Park Street Stafford ST17 4AF Tel: (0785) 47226	Distributed intelligence and micro-computer based centralised energy management systems
Chillitrol Service First International 48 Southwark Street London SE1 Tel: 01-403 1688	Dedicated air-conditioning control
Crown House Engineering Ltd (Design & Installation) Crown House, 28 George Street Balsall Heath Birmingham B12 9SB Tel: 021-440 3641	Centralised comprehensive energy management systems (micro-computer based) Distributed intelligence and micro-computer based centralised energy management systems Boiler control
Dem Controls Nunsmere Hall, Tarporley Road Sandiway, Northwich Cheshire CW8 2ES Tel: (0606) 888223	Dedicated lighting control
Drayton Controls Ltd West Drayton Middlesex UB7 7SP Tel: (0895) 444012	Distributed intelligence and micro-computer based centralised energy management systems
Energy Conservation Systems Ltd Gresham House Twickenham Road Feltham Middlesex TW13 6HA Tel: 01-898 9931/2/3	Dedicated lighting control
Energy Technology Mercia Way Bells Close Industrial Estate Newcastle upon Tyne N15 6UF Tel: (0632) 644744	Distributed intelligence and micro-computer based centralised energy management systems
Energy Technology & Control Ltd 25 North Street Lewes East Sussex BN7 2PE Tel: (079 16) 6101	Boiler control

Manufacturer/supplier	*Activity(ies)*
Floorplan Electrical Ltd Unit 7, Strawberry Lane Willenhall West Midlands WV13 3RS Tel: (0902) 61469	Dedicated lighting control
Freeman Enercon Ltd Science Park Milton Road Cambridge CB4 4BH Tel: (0223) 315432	Distributed intelligence and micro-computer based centralised energy managment systems
GEC Meters Stonefield Works Stone Staffs ST15 0RS Tel: (0785) 812111	Dedicated lighting control
Gent Ltd Temple Road Leicester LE5 4JF Tel: (0533) 730251	Dedicated heating control Distributed intelligence and micro-computer based centralised energy management systems
Home Automation Ltd Pindar Road Hoddesdon Herts EN11 0ET Tel: (0992) 460355	Dedicated lighting control
Honeywell Control Systems Ltd Honeywell House, Charles Square Bracknell Berks RG12 1EB Tel: (0344) 24555	Dedicated heating control Centralised comprehensive energy management systems (micro-computer based)
Horstmann Eng Products Ltd Newbridge Works Bath BA1 3EF Tel: (0225) 21141	Dedicated heating control
Information Transmission Ltd Unit 1, Bone Lane Newbury Berks RG14 5PF Tel: (0635) 35558	Distributed intelligence and micro-computer based centralised energy management systems
Invicta Energy Management Ltd Units 127/128 Ellingham Ind Est Ashford Kent Tel: (0233) 38344/38345	Distributed intelligence and micro-computer based centralised energy management systems

Manufacturer/supplier	*Activity(ies)*
ISS Clorius Ltd Worton Drive Reading Berks RG2 0TG Tel: (0734) 75200	Centralised comprehensive energy management systems (micro-computer based) Distributed intelligence and micro-computer based centralised energy management systems
JEL Energy Conservation Serv Ltd Bramhall Moor Lane Ind Park Pepper Road Hazel Grove Stockport SK7 5BW Tel: 061-487 1111	Dedicated heating control Distributed intelligence and micro-computer based centralised energy management systems
Johnson Controls PO Box 79, Stonehill Green Westlea Down Swindon Wilts SN5 7DD Tel: (0793) 26141	Dedicated heating control Centralised comprehensive energy management systems (micro-computer based) Distributed intelligence and micro-computer based centralised energy management systems
Landis & Gyr Billman Ltd Victoria Road North Acton London W3 Tel: 01-992 5311	Dedicated heating control Centralised comprehensive energy management systems (micro-computer based)
Margaux Controls Ltd Rusham House, Rusham Road Egham Surrey TW20 9SL Tel: (0784) 31101	Distributed intelligence and micro-computer based centralised energy management systems
Micro Energy Management Systems The Birches, Woburn Lane Apsley Guise Milton Keynes MK17 8JR Tel: (0908) 583801	Centralised comprehensive energy management systems (micro-computer based) Distributed intelligence and micro-computer based centralised energy management systems
Macro Systems Ltd 17 Claremont Crescent Edinburgh EH7 4HX Tel: 031-557 2623	Distributed intelligence and micro-computer based centralised energy management systems
Myson Brooks Ltd Vulcan Way, New Addington Lodge Hill Croydon, Surrey Tel: (0689) 41080	Distributed intelligence and micro-computer based centralised energy management systems
NEI Electronics Kingsway Team Valley Gateshead Tyne & Wear NE11 0QJ Tel: (0632) 870811	Distributed intelligence and micro-computer based centralised energy management systems

Manufacturer/supplier	*Activity(ies)*
Osram (GEC) Ltd PO Box 17, East Lane Wembley Middlesex HA9 7PG Tel: 01-904 4321	Dedicated lighting control
Panelwire Controls Abbott Street Heanor, Derby Tel: (0773) 75646/66348	Dedicated heating control
Rapaway Eurohurst House, Oakenshaw Road Shirley, Solihull West Midlands Tel: 021-745 3144	Distributed intelligence and micro-computer based centralised energy management systems
Richwest Electronics PO Box 7 Hartlepool Cleveland TS24 0RD Tel: (0429) 67831	Distributed intelligence and micro-computer based centralised energy management systems
Robertshaw Controls Ltd Raans Road Amersham Bucks Tel: (024 03) 22311	Distributed intelligence and micro-computer based centralised energy management systems
Satchwell Control Systems PO Box 57, Farnham Road Slough Berks SL1 4UH Tel: (0753) 23961	Dedicated heating control Distributed intelligence and micro-computer based centralised energy management systems
Sauter Automation Ltd 165 Bath Road Slough Berks SL1 4AA Tel: (0753) 39221	Centralised comprehensive energy management systems (micro-computer based) Distributed intelligence and micro-computer based centralised energy management systems
Seaboard Energy Systems Antrobus House, 18 College Street Petersfield Hants GU31 4AD Tel: (0730) 66222	Centralised comprehensive energy management systems (micro-computer based)
Simplex GE PO Box 102, Ash Hall Stoke on Trent Tel: 078-130 3003	Dedicated lighting control

Manufacturer/supplier	*Activity(ies)*
Staefa Control Systems (UK) Ltd Staefa House, Moore Lane Staines Middlesex TW18 4XW Tel: (0784) 61616	Centralised comprehensive energy management systems (micro-computer based)
Systemation Ltd Systemation House, Pond Road Shoreham by Sea Sussex BN4 5WU Tel: (0791) 73957	Distributed intelligence and micro-computer based centralised energy management systems
TA Controls Ltd LEA Industrial Estate Lower Luton Road Harpenden Herts Tel: (058 27) 67991	Centralised comprehensive energy management systems (micro-computer based) Distributed intelligence and micro-computer based centralised energy management systems
Technology Ops & Resources Ltd Unit 2, Sandbank Industrial Estate Sandbank, Dunoon Argyll PA23 8PW Tel: (0369) 85561	Distributed intelligence and micro-computer based centralised energy management systems
Taylor Instruments Ltd Crowborough Sussex TN6 3DU Tel: (089 26) 2181	Boiler control
Transmitton Ltd Smisby Road Ashby-de-la-Zouch Leicester LE6 5UG Tel: (0530) 415941	Centralised comprehensive energy management systems (micro-computer based) Distributed intelligence and micro-computer based centralised energy management systems
Trend Controls Station Road Horsham West Sussex RH13 5CU Tel: (0403) 69612	Distributed intelligence and micro-computer based centralised energy management systems
Watkins-Elliot Sys Engineers Ltd 123 Catford Hill London SE6 4PR	Dedicated air-conditioning control
Westinghouse Electric Haden House, Argyle Way Stevenage Herts Tel: (0438) 726177	Boiler control

Appendix II

THE ENERGY CONSERVATION DEMONSTRATION PROJECTS SCHEME

The Aims of the Scheme

The Energy Conservation Demonstration Projects Scheme (ECDPS) came into operation in April 1978. It is sponsored by the Energy Efficiency Office of the Department of Energy.

The purpose of the scheme is to stimulate investment in new ways of using energy more efficiently. It works in two ways. Firstly, companies get grants to mount demonstrations; this accelerates the rate at which novel projects occur. Secondly, information from the demonstrations is disseminated to other companies; this stimulates the 'replication' of successful projects, thus accelerating the overall rate of investment.

A demonstration project is a full-scale trial of a piece of fully-developed equipment or process, under normal working conditions, with the aim of establishing its technical and economic viability. The equipment need not be novel; often it is the application that is novel, or sometimes several well-established items are combined in a novel way so that it is the whole system which is demonstrated and evaluated.

A demonstration is not a laboratory trial, nor even a pilot scale production unit. What works in the laboratory or at a pilot scale does not always prove satisfactory under full scale working conditions. It is this last hurdle between development and successful commercial application that the ECDPS seeks to overcome.

Parties to a Demonstration Project

A typical buildings demonstration project supported under the ECDPS will normally involve four principal parties, namely:

- a host organisation, who receives the ECDPS grant to help offset the risk of failure
- the equipment supplier, who will stand to gain from a successful demonstration through the marketing benefits he will receive

–a monitoring organisation appointed by ETSU, who carries out an independent assessment on behalf of government and who provides the necessary objective reports

–government, who needs to preserve the national interest by ensuring that all viable new energy conservation technologies are taken up to the maximum extent

and occasionally:

–the energy supply industries (oil, coal, electricity and gas, etc), who might have an interest in the results and become involved in the project.

Independent Assessment

This is a crucial element of a demonstration project because it is the principal means by which information is generated, hence the need for 100 per cent government funding in order to ensure high standards and objectivity. The principal aims of monitoring under the ECDPS may be summarised as providing answers to the following questions:

–what are the energy savings and how have they been made?

–what are the consequent energy cost savings?

–what are the other (non-energy) benefits, if any?

–what is the overall cost-effectiveness of the installation?

–what practical lessons arise from the project that might enable future users of similar systems to make the best possible use of them?

–what other lessons are there for the future; should future projects be tackled in a different way? How could the equipment be improved?

With demonstration projects involving improved control systems in buildings, monitoring contracts are placed with organisations that have a reasonable combination of skills and experience in building services, energy auditing, microprocessor-based monitoring and control systems and determination of energy savings due to specific efficiency improvement measures. This is a demanding specification which, to some extent, is breaking new ground. Firms chosen to do this work have been drawn from university research departments, government laboratories, research associations, large building services consultancies and small specialist consultancies. They have demonstrated different strengths and weaknesses and all have been on steeply rising learning curves.

Criteria for ECDPS Support

The ECDPS is essentially a discretionary scheme; projects are selected for support only if they are likely to constitute a worthwhile practical demonstration in an area where the need for such a demonstration has been previously identified. The principal criteria for support under the scheme are:

(i) The technology proposed to be demonstrated needs to be novel, either in itself or in the way it is proposed to be used. For example, energy management technology is no longer completely novel in itself, but the use of microprocessor-based electronic monitoring and control systems in particular building sectors, such as schools, hotels, retail chain stores, etc, could be sufficiently novel to meet this particular criterion.

(ii) There must be a substantial and realistic potential for achieving national energy savings, i.e. of at least 10,000 tce per annum, through replication of the project throughout its market sector.

(iii) The proposed installation needs to be viable within its particular market sector. That is to say, it must offer a prospective return on investment that is acceptable when judged against the normal investment criteria of that sector. For example, crude payback times of under one year might be demanded by the hard-pressed industrial sector; one to three years would probably be acceptable to other commercial organisations, while three to five years would satisfy most local and health authorities.

(iv) The project should be based on British equipment, where suitable British equipment is available. Foreign equipment can be used where no appropriate British equivalent exists, each case being judged on its own merits.

(v) The proposed demonstration project must represent cost-effective use of the government funds allocated to it. The yardstick applied is that the annual value of the long-term target national energy savings expected to be achieved through replication should be at least twenty times the total cost to government of the grant plus the independent monitoring exercise.

(vi) The degree of risk involved in the project must be of the right order. Projects with minimal risk do not normally need demonstration project scheme support. Projects with an excessive risk of technical or financial failure cannot be supported with public money. It is those projects which are believed likely to succeed, but where there is an intermediate level of uncertainty and risk, that the scheme is aimed at.

(vii) Sometimes, there is an element of R & D needed before the project can be fully implemented. In other cases, an initial feasibility study is required. The need for these preliminary studies is also taken into account in assessing the case for support of individual projects.

Project Approval and Management Procedures

ECDPS projects are managed by teams of project officers at the Energy Technology Support Unit (ETSU) of the Department of Energy. Each Project Officer is responsible for a particular sector of energy use (e.g. iron and steel making, cement and pottery, public sector buildings, etc) or energy conservation technology (e.g. electronic energy management systems, heat pumps, etc). ETSU is based at the Atomic Energy Research Establishment of the United Kingdom Atomic Energy Authority (UKAEA) located at Harwell in Oxfordshire. The setting up of this unit forms part of the diversification of the UKAEA away from its principal historical role of conducting underlying R & D to support the nuclear industry and into non-nuclear and contract R & D work. A small sister team (BRECSU) at the Buildings Research Establishment near Watford deals with demonstration projects in the domestic housing sector.

The project officers are responsible for the determination of strategies for support of projects under the ECDPS in their various areas of technical or sector responsibility. They also carry out assessments of initial proposals from prospective demonstration host organisations before sponsoring these through the formal technical and financial assessment procedures of the ECDPS. If a project receives ECDPS support, the Project Officers ensure that the government interest in them is properly managed. The principal milestones of the life of a typical project are:

(i) Preliminary assessment at ETSU against the underlying criteria and strategy for support.

(ii) Overseeing by ETSU of an appropriate independent monitoring program by a suitably qualified organisation.

(iii) Submission of complete project details for formal assessment and recommendation by a government steering committee (the ENCORD Programme Committee). This comprises officials of government departments concerned with energy conservation (i.e. Departments of Energy, Trade and Industry, Environment, Agriculture, Fisheries and Food) plus independent members drawn from senior (technical director) level from major British industrial and commercial concerns (e.g. John Laing, Pilkington Glass, Blue Circle Cement, etc).

(iv) Processing the formal financial approval by the UKAEA and Department of Energy of the costs of the grant and monitoring for projects recommended by the committee for support.

(v) Drafting and negotiation of separate contracts between the UKAEA and the host and monitoring organisations.

(vi) Execution of a period of monitoring prior to the commissioning of the energy conservation measure (where possible) in order to establish a baseline of energy efficiency against which subsequent savings can be judged.

(vii) Similarly, a corresponding period of monitoring following the commissioning of the installation.

(viii) Reporting by the monitoring organisation of the results of analysis and interpretation of the monitoring data.

(ix) Publication of the results by ETSU backed up by special publicity and promotion events such as exhibitions, open days, seminars, etc designed to maximise the impact of the demonstration and levels of replication subsequently achieved.

Fuller details and background to the ECDPS are given in two publications: 'Guidance to Applicants' and 'The ECDPS–What you Need to Know' which are available (free), together with copies of brief profiles on individual projects and published monitoring reports from The ECDPS Enquiries Bureau, Energy Technology Support Unit, Atomic Energy Research Establishment, Harwell, Didcot, Oxon OX11 0RA, Telephone (0235) 834621.

INDEX